This book is dedicated to all those perfect leaves of grass,
who have already gone to the dirt to grow.

Contents

Introduction: Nothing (but Flowers) 9

Chapter 1: Papa Was a Rolling Stone 15

Chapter 2: Feed Me (Seymour) 35

Chapter 3: What's New Pussycat? 57

Chapter 4: Revolution 77

Chapter 5: Power of Equality 93

Chapter 6: Oops Upside Your Head 117

Chapter 7: Under My Thumb 143

Chapter 8: War! What Is It Good For? 169

Chapter 9: Under Pressure 189

Chapter 10: Bring Out Your Dead 209

Chapter 11: Tainted Love 229

Chapter 12: Take This Job and Shove It 253

Chapter 13: Panic ... 279

Conclusion: Karma Police 299

Acknowledgements: Some of My Friends 313

Index 315

INTRODUCTION

Nothing (but Flowers)

The amount of time humans have spent living in cities is an infinitesimal speck in the scope of hominid evolution. It took our species something like 200,000 years to get around to trying out living in the same place all year round. Then, it took thousands of years of experimentation for those patchy early settlements to become the cities we recognise today, and it's just in these last few years that the number of city dwellers has finally outstripped that of our country cousins. We are now officially an urban planet – but why did it take us so long to get here? If cities are so great, why are they full of things that kill us? Urban life serves up a terrifying cocktail of the most dangerous things known to our species – disease, inequality and, of course, other people. It's not unreasonable to ask: why have we made cities this way?

However, there is a better question, and it's one that you can only ask with a bit of patience and a whole lot of dead people on your hands. If we look at the lives and deaths of people through all the different experimental stages of urban life, we can start to see some very interesting patterns in these urban pioneers. Patterns of disease. Patterns of malnutrition. Shrinking faces and growing numbers. Broken backs, broken skulls, bone missing where you want it and piling up where you don't – the speechless generations of the past still have quite a story to tell. It's through the skeletal remains of the city dwellers of the past that we can answer the question this book asks, and it matters to everyone alive today: why have cities made *us* this way?

It's fairly simple to look out of a twenty-first-century window at a sky full of hazy smog, over the glare of fossil-fuel-powered lights and particulate-belching vehicles, and come to the conclusion that cities are a hazard to human health. Are we really prepared to declare that the barely 2 per cent of time our species has spent on earth experimenting with cities is the way of the future, or are our current urban lives just one more pit stop along the long trail of hominin* wandering? Weren't we happier then?

Look at them. Just there. Coming over the ridge, at first silhouetted starkly against the rising sun but with the details of their bodies edging into visibility as they break the top of the dune and begin their descent. He (and we must always talk about him first because androcentrism is a palaeo tradition) is a paragon of human virtue. He is whippet-lean, with tight bulging calves and washboard abs. His hair, depending on your personal taste, might be in a man-bun, with sun-kissed flyaway strands held back by some sort of musk ox leather hairband; or maybe he's got neat dreads that really frame his face. She, on the other hand, is more difficult to describe, largely because she's got children strapped to most of her body, covering the bits that the weirdly tiny and ragged scrapes of animal hide she's wearing don't, and is carrying yet another infant in her admittedly well-muscled arms. He, by contrast, gets a spear and one of those shell necklaces surfer guys wear. This, ladies and gentlemen, is the wave of the future past. Lean, low-fat bodies, natural-fibre clothing and full-attachment parenting. Our model couple, meandering across the dunes of my advertising-and-Raquel-Welch-blighted imagination, are the absolute apogee of what our modern neuroses tell us life was like when we lived 'naturally'. Those washboard abs (his and hers, which you can see if she detaches the middle-sized child temporarily) are the product of eating a 'palaeo-diet', those enviable leg muscles the product of 'barefoot' exercise, and the brood of happy children the product of (and yes, I swear to you now, this is an actual thing) 'palaeo-parenting'.

What a lovely image. How far removed from the depressing sight of yet another overweight family group fighting in

---

* Hominins are us and everything more closely related to us than to chimps. This is for the best really, given what chimps are like.

their car outside a drive-through fast food restaurant. The proponents of the palaeo-diet tell us that, if those harassed and harried parents in the leased minivan just gave up the burger bun and went for the 'animal style' option, his incipient diabetes and her chronic gut issues would simply disappear, and they'd both be left with nothing but toned muscles, glowing skin and glossy hair. Palaeo-lifestyle advocates, like those who reject shoes or letting your toddler take his attention issues out on an iPad rather than your frayed last nerve, have similar advice. It comes from a thousand corners of the internet, magazines, TV and talk shows – the not-so-subtle message that we've all become a little too modern for our own good. Too settled, too sedentary, too dependent on our easy carbs and cushioned walkwear.

But are they *right*? If we can look around at how we live now and see that, in many ways, it's killing us – well then, why are we doing it? Why do we live in our big cities, with our Big Gulps and our bigger bellies? There's a quick answer and a longer one. This book, I hope, is the longer one, but we can get a short answer just by zooming in on our palaeo-couple as they trudge across the savanna. What do we really know about them? We can imagine what we like, but how can we get real information on what it was like to live as a hunter-gatherer? It's actually not that hard. Archaeology, as most people will know, is the science of recreating the human past. *Bio*archaeology, which most people ought to be forgiven for never having heard of, is the science of reconstructing lives in that human past. Bioarchaeology takes the remains of living things, such as skeletons and teeth (but even potentially hair, skin and even nails), and from the wealth of information locked inside the structure, composition, shape and size of these ancient clues, tells us how people (and animals) lived and died in the past. Human remains can tell us about life in the past to an unbelievable level of detail. For instance, let's look back at our palaeo-parents.

Take a closer look at Spear Guy. How tall is he? Actually, fairly tall – around 1.79 metres, or about 5 foot 9 inches. When archaeologists find his bones in 15,000 years' time, they will take careful measurements of all of his limb bones – the femur, the humerus, the tibia – and will plug these measurements into a formula that will estimate his height. Those musclebound limbs we've been seedily admiring will also show up, but this time with the help of a machine. By CT-scanning his bones, bioarchaeologists can reconstruct his bone density, giving us a clue as to whether he was more couch potato or rolling stone. On the outside of his bones, more clues about activity abound. Take a look at his arms, especially the one he holds the spear in. Where his most pronounced musculature is in life, so will it be in death – the places on his arms where those spear-thrusting or throwing muscles attach will be more marked than on his gender-stereotyped partner, who presumably isn't stabbing or throwing the kids she's carrying in quite the same way as she would a spear.

Now, let's take a look at his teeth. Uff. Well, here's where the palaeo-diet folks might want to start looking around for another model. Our friend here has certainly got teeth, and while they're not the 32 your modern dentist would expect, because a few of them have fallen out by now, they're not doing so bad. Of course, they're not very pearly white – they're not very anything: the big chewing teeth in the back have been worn nearly flat, and his front teeth have been deliberately pulled out sometime in his teens – because even hunter-gatherers have insane fashion trends. That wear on his teeth tells us that he's no longer a spring chicken, something we find further proof of as we delve into the bony remnants of his vertebrae, his elbows and his knees. Joint disease was creeping up on Spear Guy, a considerable achievement considering that human life expectancy in the Palaeolithic period we are considering was closer to 40 than 100 and many of his contemporaries would have lived and died without having had the time to develop age-related illnesses.

Well, Spear Guy may have some issues, but how about Raquel there? Her bones and teeth tell a similar story of an active, wearing lifestyle, but it's the three kids she's hauling that really impress. And well they should: modern hunter-gatherers usually have children spaced fairly well apart. The reasons for this aren't quite so straightforward, but allow Raquel to demonstrate the basic principle by attempting to feed infant C while toddler B is strapped to her chest and child A is grabbing her hand and periodically trying to climb her legs like she's some sort of Ent. Arranging the sling holding the bulk of toddler B to one side, she attempts to bring infant C closer, but at the same time child A makes an aggressive landing on her left leg and begins to scale the side of her body currently weighted down with fidgeting toddler B. There is a brief and glorious moment when her taut, tensed muscles appear as though they might take this final strain, calves bunched against the weight and arms straining to hold infant C to her chest. But alas, the whole breeding edifice comes tumbling down, sending mother and children A through C headfirst into the nearest sand dune for maximal comedic effect. There are those among you with better sensibilities than the illustrators of most 'palaeo' books who might ask why Spear Guy didn't, you know, put down the spear and carry the baby for a bit or something … but obviously that would spoil the image.

The point I'm laboriously trying to make here is that what we think of when we envision the human past is partly a myth that tells more about where we think we're going wrong with our own lives today than anything that happened thousands and thousands of years ago. Marlene Zuk, a behavioural ecologist and evolutionary biologist, wrote a brilliant exposé on 'palaeo-fantasy' in a book of the same name, which discusses exactly this point. But what we're interested in here isn't just a takedown of some ridiculous celebrity-fuelled diet or fashion trend. What we want to know is: what happened to Spear Guy and Raquel? Why did they decide to give up on the big savanna and start spending their time over at the river's edge, swapping a never-ending horizon for the more manageable commute associated with settled life? And what did that decision do to his leg muscles or her baby-juggling skills? Because we can look at the couple eating fast food in their car in a nameless city, success stories and paragons of human achievement in terms of the race to urban living that has characterised the last 15,000 or so years of human history, and we can look at the legions and legions of our ancestors who died along the way to that fast food parking lot, and see, locked away in the cells and structures of bones and teeth and hair and skin, that we haven't just built cities. Cities have built us.

This book looks at human adaptation in the face of human invention. It also tries to unpick some of the factors that underlie our urban lives, to see which might be killing us and which might not be so bad. Some of the things we know about cities in the modern world can be traced right back to their beginnings, like the role that inequality plays in determining who dies in a slum and who gets top-flight medical care. The story of humans in cities is, if you like, a kind of micro-evolutionary tale, one that we can read if we follow Monty Python's advice and start bringing out the dead. With bioarchaeology, we have the unique opportunity to get a very inside look at what the move from savanna to city has done to our bodies and our health, from slightly before the very beginning some 15,000 years ago until the Industrial Revolution and the start of our modern age.

CHAPTER ONE

Papa Was a Rolling Stone

The valleys of Jordan are a supremely arresting geography. Great rust-red outcrops of raw geology are up-thrust into a mellow blue sky, riddled with cracks you can just about drive a Toyota Hilux through. The edge of the Jordan Rift Valley forms a forbidding wall that occupies the entire eastern horizon as you drive south along the Dead Sea Highway from Amman, slowly encroaching on the flat, salted-earth badlands, riddled with unexploded mines left in the aftermath of the Arab–Israeli wars, and on towards Wadi Arabah. The closer the highway presses against the side of the massif, the more obvious the small cracks in the mountain landscape become. These are channels and rivulets worn by water, time and tectonics that provide the only access for Bedouin, camel or truck into the hidden world of the valleys beyond. Because, although the bleached yellow-grey of sparsely vegetated rocky hillsides and the bitter salt spread of the Dead Sea shore suggest an unforgiving, inhospitable landscape, the truth is that this desert country is home to one of the earliest experiments in what was to become a major human revolution: settling down.

I had plenty of time to make these observations in January of 2012, wedged firmly into the jump seat of a reasonably late-model pickup, one of three four-by-four vehicles borrowed from the Council for British Research in the Levant. At the wheel was the regional director of the Council, Bill Finlayson, who had organised a sightseeing trip for a ragtag bunch of archaeologists participating in a conference session on the Neolithic in the Near East during the seventh annual World Archaeological Congress. I'd left my own rental car in the hotel car park and was packed in with some of the most knowledgeable experts in the region – academics whose careers are dedicated to unpicking the secrets of the human transition from hunter-gather lifestyles to settling and living, year-round, in a single place. We'd had a fantastic session discussing details and discoveries, even though everything that year carried a sombre undertone as the situation in Syria finally unravelled. A 15-minute report on a site in Syria turned into a heart-wrenching 30-minute discussion about the safety of the site workers and last-heard-of's. Everywhere in Jordan that year, white UN refugee tents had been popping up like mushrooms: the 200,000-plus seasonal workers from Syria who normally came south to work in Jordanian fields had refused to return home, and their tents had taken on semi-permanent, lean-to aspects – a strangely nostalgic nomadic trajectory in a country that spent years trying to get the Bedouin out of their traditional tents and into settled villages.

This slow transition of the no-longer-seasonal workers from temporary tents to more sturdy semi-permanent shacks and huts is a very modern phenomenon, but it does strangely mirror the very process we were all driving through Jordan to understand: the act of staying put. As we moved south down the Dead Sea Highway in our convoy of mismatched trucks, Bill, who has worked in Jordan off and on for most of his professional life, explained the unexpected geopolitical repercussions of Jordan's booming produce sector and Syria's burgeoning civil war. It struck me as a strange juxtaposition. Here we were, setting off on a tour of some of the very earliest permanent settlements known to humankind, all while driving through a landscape littered with signs of twenty-first-century nomads, both Bedouin and migrant workers. Looking at the forbidding landscape, you'd be forgiven for wondering why anyone would want to settle here, of all places.

It's a question that only became more pressing as the cars turned east onto the flat belly of one of the many little inlets into the rocky foothills, about halfway down the highway towards the Red Sea, and started rumbling gently over the grit and sand that blew incessantly across the snaking strip of paved road. Facing away from the sea at last, with the mountains temporarily held back by the intervening basin, the vista was of yellowish gravel and sand interspersed with occasional patches of date palm, cradled by rapidly rising purple-grey foothills. An extraordinarily photogenic camel and her progeny, a gangly fellow about as tall as the Toyota and roughly 80 per cent knees, watched expressionlessly as we meticulously followed the hairpin curves of the road as it twisted and turned across the flat, featureless valley floor. I wondered (vocally) at the time, wedged as I was with my knees up against the bench seat and with Mihriban Özbaşaran and Güneş Duru, director and co-director respectively of the excavations at Aşıklı Höyük in Turkey,* squashed in next to me, why there was quite so much meandering involved in traversing such a flat plain. This, it turns out, was a very foolish question. No sooner had we passed what I can only describe as a Bedouin car boot sale, with the cars in question actually small pickups loaded with sheep or serving as hitching posts for bored-looking camels, that Bill brought the whole convoy to a gravelly halt.

Out we all piled, a reverse clown-car manoeuver with more North Face than face paint, considering the high field-archaeologist quotient we were hauling, to confront the obstacle. Here, camouflaged by the invariable beige-ness of the landscape, was an abrupt and unarguable hole in the road. The hole was actually more of a ditch, if you'll accept something around half a metre (2 feet) deep and 5 metres (16 feet) wide as a ditch. Bill stared at it. We stared at it. The four-by-fours behind us all stopped, emptied, and the rest of the group gathered around and stared at it. Slowly, I realised that what I was looking at wasn't just an inconvenient hole randomly targeting our road: this was the very nature of the valley floor. The valley we'd been driving through was absolutely riddled with the aftermath of the violent, earth-carving torrents of water that stream down from the hills and

* Güneş Duru is also the lead guitarist in the Turkish rock band Redd, which suggests that either archaeology or rock music in Turkey is underpaid.

fill these dry, rocky valleys with life. Snow from high up on the Jordanian plateau trickles down into rivulets, which build into streams as the water comes crashing down, careening through a successive series of canyon valleys. These flash floods might wash out roads every year, but they are key to letting life flourish in this hard, rocky landscape.

These valleys are called '*wadi*' in Arabic, and they form the cornerstone of an ecosystem that held a unique appeal for humans thousands of years ago. The wadis provide a mix of environments to exploit, and the Jordanians of prehistory were canny enough to realise this. They exploited not only easy access to water, but also easy access to the many animals that came for the water. Water flowing down off the high Jordan plateau would have run down into the arid valleys, briefly bringing a flourish of vegetation and life totally absent on a January morning. Haunting the edges of different ecozones where different resources can be found is a traditional hunter-gather strategy, to position yourself for ease and opportunity where different types of resources meet – for instance, the edge of a river, situated between uplands and lowlands. The confluence of water has actually been a feature of many early experiments in sedentism – one could point to the Cahokia Mound Builders of the Mississippi River, or the Shangshan site on China's Yangtze River. All cities need a river. But Bill was taking us to see something else, something new; well, something thousands and thousands of years old, but a revolution at the time.

First, however, we needed pickaxes. Approximately equal numbers of senior academics hacked away at the impromptu obstacle as advised from afar on how best to hack away. This saw us clear the first hurdle in under 30 minutes, after which the washouts were all surmountable by the much-abused four-by-fours. We managed to cross about half of the valley floor before coming up on a new-build Bedouin village, a neatly organised grid of dusty pink and yellow houses. Here we were waylaid by colleagues and friends of Bill's, and I learned that what I think of as coffee is nothing like actual coffee – at least the way the Bedouin make it. Bedouin coffee is a weak greenish liquid, served scalding hot and drunk in a rapid, throat-burning gulp from a tiny cup about the size of a thimble, which you are meant to drain in one and then pass on to the next person. The flavour is immense, however – the coffee itself is made from fresh rather than dried beans, and like everything else in Jordanian cuisine, has cardamom added to give it a heady, spicy kick. There is an art to drinking anything that hot, and I'm not sure I managed with any grace, but at least we got through the valley and up to the foothills themselves without causing mortal offence.

After leaving the relatively hospitable valley floor, with its adorable baby camels and occasional patches of greenery poking through the grey rock, we drove up Wadi Faynan, home of Bedouin coffee pushers.* Finally, we managed to ford a number of water-free streambeds to arrive at the important early site of Wadi Faynan 16 at a slightly higher elevation, overlooking the main canyon. This was one of those early experimental villages, dating back 11 millennia, and it is impossible to describe how bleak and desolate an environment that forlorn hilltop overlooking the wadi appeared in mid-January. The ground is a sort of uniform grey, and the only colour comes from the minerals in the wadi sides as they descend down to patchy greenery clinging to the path of the springtime floods on the valley floor below. But here, we have good evidence of year-round inhabitation of the little round houses dug halfway into the earth and bolstered up with mud brick. The Wadi Faynan 17 archaeological project has actually reconstructed one of these ancient houses, and we all had the opportunity to crawl down a narrow defile cut into the earth, ramping about 1.5 metres (5 feet) down into a round inner room with cold mud-brick walls and a low straw ceiling overhead. It didn't fit many people at once (in fact, someone with a long reach could probably have simultaneously touched opposing walls),

* I came to realise that there are very few Jordanians who do not appear to have made it their mission in life to introduce you to delicious coffee, up to and including bored military personnel on the midnight desert road shift who invite you to have coffee with them in the tank they are using as a windbreak. Thanks guys!

and would have been a health and safety nightmare with a hearth fire going. This is one of the advantages of experimental archaeology: pointing out obvious flaws in our imaginings of the past by, literally, smoking them out.* Currently, the site interpretation seems to be that fires, cooking, grain-grinding, *etc.* would have happened outside; the big grinding stones used to mash figs, pistachios and wheat into edibility are found outside these little shelters. It seems that this early experiment in settling down would actually not have looked too different from the seasonal occupations that have been found in this area before, with lots of communal activity taking place in a shared, open environment. The exciting difference at Wadi Faynan 17 is that this shared social space was being broken up into smaller 'house' units that, instead of lasting a night or a season, seem to have been occupied all year round.

As the four-by-fours dropped into low gear and we began our slow, steady climb up the side of the wadi, there was plenty of time to think about the people who had lived here, in this beautiful but harsh land, millennia ago. For long periods in the human past, this area was sparsely inhabited by fairly mobile hunter-gatherer groups. About 12,000 years ago in the Levant, which encompasses the fertile landward part of the Eastern Mediterranean, the first archaeological evidence of longer-stay human occupation patterns begins to emerge. The Natufian culture of the region seems to settle itself into the landscape a bit more permanently, building drystone-walled shelters in and around cave systems, burying its dead in recognisable cemeteries and installing the ancient equivalent of large, immobile white goods: grinding stones and storage pits. Some 3,000 years before the development of agriculture, its seasonal camps started to take on a more permanent feel, and even the tiny bones of house mice have been used to suggest a more long-term occupation strategy.

* I am reliably informed that this experiment actually took place. No archaeologists were (permanently) harmed.

After a millennium or two, however, the world these groups inhabited began to shift around them as the climate slowly turned over the course of a thousand-year spell of global cooling known as the Younger Dryas. Vegetation changed, animals changed and the resources that human life depended on changed across the landscape. The Natufian cultural markers become associated with less permanent-looking settlements once more, and the early period of sedentism seems to evaporate as a failed experiment. While no one agrees exactly how much this shift impacted human behaviour, or how settled the early Natufians really were, there are fairly clear archaeological indications that as climate shifted, there were real changes in the way people lived.

After many millennia of seeing intense hunter-gatherer activity, increased rainfall seems to coincide with the emptying of the Wadi Faynan area during the Pre-Pottery Neolithic A. It isn't until the Pre-Pottery Neolithic B between roughly 10,500 and 9,500 years ago, a more arid period, that we see the thin scars of temporary seasonal encampments that littered the wadi replaced by the heavier footprints of solid buildings. These were occupied all year round, as we can see from the materials and effort that have gone into their construction, as well as the evidence from the small finds. There are a lot of fire pits with a lot of radiocarbon-dateable charcoal, as well as a wealth of animal bones, lithic tools and even tiny micro-artefacts like the pollen spores of long-dead plants. All of these tell a story of a year-round settled people. Looking at the bones of the animals they killed and the residue of the food they ate, it becomes apparent that the newly settled groups were still relying on hunting and gathering strategies for subsistence. Specialist analysis can reveal if the shape of a tiny flake of plant remains comes from a wild or domesticated variety, and the same is true of animals – more on which in later chapters. The final change from the Mesolithic world of long-ranging mobile groups to the stationary Neolithic can be seen in the people themselves, or rather, in their numbers: settlements hold not the tens of people seen in modern-day hunter-gatherer groups, but hundreds and possibly thousands of people. But what possessed all those people to hunker down and stop wandering the landscape? And where did they all come from?

We can start to understand this early experiment in 'staying in' by looking at what it means, physically, to be a hunter-gatherer. Take everything you've ever heard or read about 'palaeo' diets, exercise routines and lifestyles in the popular press, and please repress it. Severely. What we do know about the Palaeolithic lifestyle has to be painfully reconstructed from rare, fragile archaeological finds – the tiny fragments of bird bones, for instance, which only recently came to light and show that pigeon was a (rather unexpected) part of Neanderthal diet on Gibraltar. Or the chemical evidence of different types of isotopes that settle into bones and teeth at different rates depending on the food you eat. We can build a wonderful picture of different proportions of meat and plant food in the diet, which only holds up until a study from just the next site over points out that your signal looks like it came from a grass-munching cow that had just digested a voracious meat-eating lion.* But if we really want to understand changes in human behaviour, we need to look at the actual humans: the evidence of individual lives and group survival … or failure. So what was hunter-gatherer life like in the austere wadis and high plateaus of Jordan?

Well, for one thing, it was probably longer. From the get-go, settling down into hamlets and villages seems to have shortened life expectancy. Trying to understand how long people lived in the past can be incredibly difficult, because the only evidence we have is … dead people. This is the fundamental challenge of the academic discipline known as palaeodemography. There are no orderly census records for us to consult to see how many people were born or died, or to judge how old they were in a given census year. All we can do is extrapolate the numbers from groups of skeletons that

* This is a facetious comment. Not to be confused with the evidence from actual faeces, which is actually incredibly useful in reconstructing diet.

happen to survive in various parts of the landscape, and I think we can all agree that a population of skeletons isn't likely to be normal.* In a normal *living* population, we see a sort of bell-shaped curve (Figure 1, below) if we lay out the whole population according to age, from birth to death. At the extremes, however, the numbers trail off because people either grow out of infancy or die out of the oldest age bracket. You wouldn't expect more babies than numbers of reproductively active females alive at any given moment in time, and the things that kill us off do tend to affect older people (*e.g.* cancers, heart disease). Sometimes the bell curve will skew younger or older, indicating that there has been a baby boom (the curve skews to the left), or some previous boom is heading towards greying out of existence (the curve skews to the right).

A palaeodemographic curve, however, is the exact opposite. It's a u-shape, with spikes at the very end of the age extremes. Does this mean that the past was entirely populated by newborns and oldies? No, of course not. What it means is that newborns and the elderly are the ones who *die*. Before

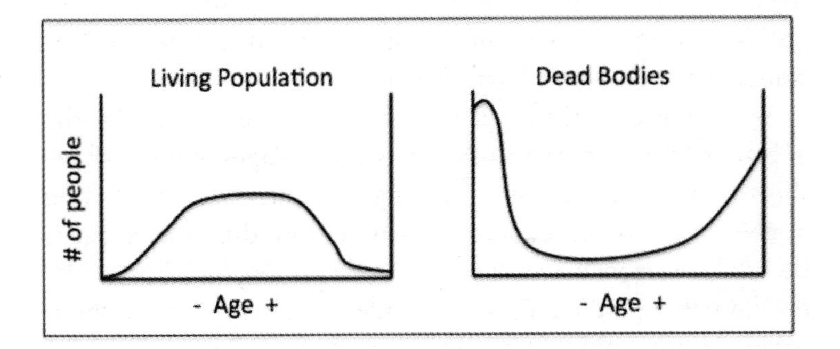

Figure 1 *Demographic curves, idealised for a living population (left) and a dead one (right).*

* This is, by far and away, the funniest joke it's possible to make in palaeodemography, and I've nicked it from Professor Tony Waldron of University College London. I'm sorry.

the advent of modern medicine, a large percentage of children died in the first four years of life. The World Health Organization still gives the mortality rate for infants under the age of one in some parts of the world as nearly 1:100. Aside from complications of birth, such as oxygen starvation, death of the mother or infection, there are a host of childhood illnesses, particularly diarrhoeal ones, that, coupled with undernutrition or lack of resources, still cause a depressingly high number of child deaths.

This brings us to the concept of mortality risk. At birth, mortality risk is very high, because there are a great many things that newborns can die of. This drops a bit after the first two weeks, then further after the first year or so, and finally reaches a sort of plateau around about four years old – or so we can estimate from a hodgepodge of census records, surveys and, of course, archaeological data. Mortality risk rises again at particular times in life; for instance, a male of draft age in a group that does a lot of fighting or a female of childbearing age in an era before antiseptics and antibiotics will both have significantly higher risks of dying than a teenager, despite what the parents of most teenagers may think. After this last peak in mortality risk, however, there is a bit of a plateau again, followed by a gentle ramping up of risk until, in the end, the curve peters out. What archaeologists uncover is this u-shaped distribution of dead people, which describes not the total number of people alive at any given time, but is rather related to mortality risk. The trick, then, is to try to work backwards from the number of infants and old folks to arrive at a reasonable approximation of the living population. It's not nearly as easy as it sounds, and already it sounds far too statistical for most people's tastes.

If you think about how a Christian church cemetery, for instance, comes to be filled, you can see the problems immediately. In many Christian traditions, children who died before baptism were not buried in the main burial ground – archaeologists might not ever see them. So, can we assume that there were no babies ever born in the entire history of the church? Again: no, of course not. We might instead turn to the ages of the folks whose bodies we do have. If the average age of burials is low, we might guess that life expectancy was also low –and hope that no one hid the bodies of all the old people (if they can hide all the babies, is it really so unreasonable to wonder whether they hid the elderly?). A low life expectancy suggests that mortality risk was high; otherwise, more people would have made it through into old age. Archaeologists examining the skeletons of hunter-gatherers and the more settled groups that followed them (or rather, refused to follow them and stopped somewhere nice) have searched for a difference between the average age at death for hunter-gatherers versus settled groups. Some of the latter seem to have lower mean ages at death than their non-settled predecessors in the same region – something that has led researchers to suggest that settled people typically suffer from *higher* mortality risk than their free-range forebears. The counter-argument runs that settled groups experienced more baby booms (on average everyone was younger), and therefore left a bigger pile of youngish skeletons behind. The sedentary trend did take off fairly quickly in archaeological terms, coming hot on the heels of a few hundred thousand years of determined meandering by our hominin ancestors. Was it all due to the birth of an immense cohort of fixed-abode evangelists?

The palaeodemographer Jean-Pierre Bocquet-Appel and colleagues have studied the age profiles of skeletons buried at sites from all areas of the world and from the periods in each region where people began to occupy the same sites year-round. They have concluded that the higher proportion of immature skeletons must be the result of more children being born (and therefore, at risk of dying). The evidence from burials also shows a more general population increase that, taken in conjunction with the number of sub-adult skeletons, suggests that the massive change in human numbers they observe across the globe during the transition to the settled Neolithic period is actually a result of increased female fertility. This increase in female fertility is seen in turn as the result of a better balance of maternal energetics – the cost of bearing and feeding an infant against the mother's own calorific needs – for sedentary women.

Having a baby is a very expensive prospect, a well-worn sentiment that has probably seen little change since our ancestors developed the ability to reflect on existence, and coming well before the advent of currency and competitive consumption of Mothercare products. The biological expense is greatest of course for the mother, as she bears the direct costs of providing enough nutrition and energy to the developing infant, both before and after birth. In situations where the mother herself does not have access to unlimited calorific resources,* she has to maintain a very delicate balance between reproduction and survival.

We primates are pregnant much longer than other mammals growing similar-sized infants: we take two to four times longer than would be predicted from the normal mammal ratio of gestation time versus body size. Of course, we also have single or very occasionally low multiple births of babies who are born with eyes and ears open, unlike, say, mother cats do. The rate of development of the foetus is slower, which means that the energy we need to devote to it is actually lower, something that could have been a very useful survival strategy in species with fluctuating food resources. Humans slow-grow even larger babies, however, than most of our nearest primate relatives: a baby gorilla will weigh about 2 per cent of its mother's body weight, and a human a whopping 6 per cent. Chimps also have babies that weigh about 6 per cent of the weight of their mothers, but interestingly, human babies clock in at roughly 15 per cent fat, whereas other primates are closer to 4 per cent. Our fat babies cost quite a lot to grow, too – it takes an estimated 77,500 calories (about 240 calories a day in a modern English sample) to carry an infant to term. Some researchers have argued that this is directly related to the relative fatness of

---

* See for instance: nutrition information, Ben & Jerry's Chunky Monkey ice cream.

human women,* but it's very difficult to extricate how much of our fat stores are dependent on modern obesogenic diets and how much on our evolutionary history; captive primates also put on considerable fat stores. Feeding the babies once they are born is another expensive prospect – the energetic 'cost' of lactation is about 500 kilocalories per day, which human mothers meet by increasing energy intake, limiting physical activity, or a combination of the two.

Bocquet-Appel has theorised that settling down is exactly what allowed humans to tip the energetic balance in favour of more babies. In primates with lower body fat stores, it takes longer for the undernourished to return to fertility, and in humans, nutritional shortfall can lead to amenorrhoea† and temporary infertility. Settling down, by this reckoning, must coincide with a net energetic gain for mothers. This assumption has been borne out by the anthropological literature to some extent: in some ethnographic examples, there is a clear trend for mobile groups to have children more widely spaced apart. For instance, the Ache people of Paraguay were, until the 1970s, highly mobile foragers. They were gradually settled over the course of a decade, but still frequently participated in foraging activities. Women with infants devoted an intense amount of energy to their offspring and spent approximately 90 per cent of foraging time in direct skin-to-skin contact with their children, never letting them get more than a metre away. By contrast, when the Ache were present at their settled home (a mission), children roamed freely and were frequently supervised by other community members or siblings. The time between the births of successive children among the Ache fell from 3.2 to 2.5 years after settling down.

The origins of the theory of the Neolithic Demographic Transition actually come from observations made by anthropologists Lewis Binford and W. J. Chasko of Inuit communities moving from a mobile to a sedentary lifestyle in

---

* Human females apparently have a higher body fat percentage than every other mammal besides the Djungarian hamster.

† When the menstrual cycle stops.

the 1970s. They found that the subsequent baby boom could be attributed to the availability of new foods and different demands on mothers' energy. Across the world, mobile groups have more widely spaced children than settled groups, averaging around four years between each birth. Generally, children are breastfed until the mother's next pregnancy, and so are weaned after at least 2.5 years. However, this is highly variable, and the difference between modern farmers and hunter-gatherers is actually not statistically significant. It should also be remembered that these observations were not made in some sort of pristine primitive vacuum, but among groups being more or less forcibly settled.

What is clear is that somehow, human populations began growing in the run-up to the Neolithic. Evidence of increasing numbers in the shape of numbers of bodies and increasing settlement size has been seen in South America, China and the Near East as each region transitioned to settled life. Many authors argue that this population boom was part of what carried the revolutionary ideas of the Neolithic – settled life and agriculture – to new places; the baby boom that followed mothers finding a chance to put their feet up meant a burgeoning population overflowing into adjacent territories. This slow ooze of Neolithic ideas is an intensely studied (and intensely contested) part of how settled life came to take over most parts of the globe. Similarly, the role of farming in the first wave of sedentary lifestyles in the human story is sufficiently complex that the next chapter in this book deals exclusively with what we know about the pros and cons of making a living from the soil. However, what is clear is that, come the start of settled life in the Natufian period beginning some 14,000 years ago in the Levant, we start to see a lot more people. Population growth kicks off in a way that floods the landscape with settlements, with burials and, of course, with bodies. This fertility revolution, unchecked, had the capacity to launch modern humans into the demographic stratosphere, after hundreds of thousands of years of more or less stable numbers. And while there was a definite boom in population, something *did* check our numbers, keeping us from the growth that could be predicted from increased female fertility. That something was, of course, the same thing that has kept its Malthusian thumb on the human population for as long as we've been around: death and disease.

The advent of settled life brought with it an increase in the risk of settled death. The problem of course, mentioned above, is that it's usually very difficult to ascertain an actual cause of death from the human remains encountered in archaeology. The vast majority of things that kill us leave no trace on the skeleton; only chronic conditions affect us long enough to produce a response in the bones that archaeologists find. Flu, measles, plague, typhoid, cholera or any of the terrible pandemic diseases that have such an effect on human populations are largely untraceable from bones alone (though we'll see in later chapters how we *can* trace them). Heart attacks, drowning, poisoning and even violent deaths might leave no trace at all. Given all this, if we want to understand what kept a lid on the fertility explosion of the Neolithic, we have to extrapolate what the overall *risk* of dying was.

There are ways to try to access this kind of information about mortality risk, embedded in the bones and teeth of people from early settled communities. For instance, locked into the shape, size and left–right balance of your skeleton are a host of indicators about health that can be teased out via bioarchaeological investigation. These physical signs can tell us about when circumstances got in the way of normal health or growth. This book will talk about them again and again, because it's our only way of getting past the silence of prehistory and all those people history didn't much bother with, and of really accessing the reality of life as a biological organism during our many adaptive escapades. So when we look to the edges of the last 10,000 years to understand whether settling down increased humans' risk of dying early, we need to start talking about physical changes from head to toe.

What is different about a hunter-gatherer's skeleton? Well, we can start from the toes on this one – or at least the lower limbs. This is the conclusion reached by researchers like Noreen von Cramon-Taubadel, who compiled a huge number of skeletal measurements of hunter-gatherer and sedentary groups from different places and times across the Near East and Europe. Her research looked at how the varying size and shape of human heads and limbs might have changed as the new trend of settling down and farming spread across Europe, pushing the hunter-gatherer lifestyle out and making way for new Neolithic settlements. You would expect to see some diversity across that much time and space (it took thousands of years for the full package of Neolithic innovations, including farming, to turn up in the remote corners of Europe), which she and colleagues duly found. But what they also reported is that, overall, hunter-gatherers had larger lower limbs than non-hunter-gatherers. They had bigger legs, even after accounting for the types of things we know have a relationship with the skeleton beyond simple genetic inheritance, like climate and latitude. Many physical anthropologists have linked the size of limbs, or their general robusticity, to patterns of activity, but understanding how activity affects the human skeleton is quite a bit more complicated than that. So while it's tempting to simply chalk up our increasingly scrawny legs to just not logging the hundreds of miles a hunter-gatherer might, other factors such as a person's ability to put on muscle (not to mention our ignorance of what the marks a lifetime of a no-longer-practised lifestyle would look like on a skeleton) prevent an easy interpretation of the data.

Skinny legs aren't the only interesting clue to life in the Neolithic that can be found by measuring skeletons, however. In the Near East, where our story starts (but also from across the globe, as sedentism and farming get going elsewhere), researchers have observed that everyone, everywhere, appears to have started shrinking once they settled down. Bioarchaeologists estimate adult height by using a series of formulas based on the long bone lengths of known-height individuals. To estimate height (or 'stature') accurately, the ideal is to have several limb-bone measurements, each with a bit of an error range, and then to hone in on a realistic guess for overall stature based on those limb-bone lengths. Even so, while the evidence of the archaeological record can be frustratingly ambiguous, it seems clear that, with the arrival of a Neolithic lifestyle, people became shorter. As this occurred thousands and thousands of years before the advent of the computer and the adoption of the hunch-shouldered, screen-staring posture so beloved by many of us today, it's difficult to know how exactly we managed to shrink. Were the populations who settled different than the hunter-gatherers, and therefore perhaps genetically predisposed to achieving different heights? Basically: did already short populations settle and then outbreed the taller populations nearby? Ancient DNA studies ('aDNA') have shown that a lot of Mesolithic groups – hunter-gatherers who lived either before or even alongside settled Neolithic groups – are genetically fairly separate from later Neolithic populations.

There are, however, other factors that can determine adult height. The big red flags here are malnutrition and other types of growth disruptions that occur during childhood, or even in the womb, but set an individual up for a lifetime. These growth disruptions actually form an important source of evidence for life in the past, and we'll come back to them in later chapters. We know from modern growth studies that stunted growth can be caused by lack of resources, particularly food. A further clue comes from the relative female- or maleness of skeletons that come from the period right after the first Early Natufian experiments in settled life. The Natufian period is broken into two sections, with the later part subject to the climatic upheaval of the Younger Dryas period. In the Late Natufian there are signs that the populations were under considerable stress, and the experiment with sedentism was called off in most regions as people returned to a more mobile way of life. As pressure on the Natufians increased, sexual dimorphism (the difference in body shape and size between males and females) also decreased, something which happens when a growing body is under enough pressure that resources that might go towards increasing dimorphism are diverted to the rather more pressing case of survival. Many of these skeletal changes are of types that can be related to developmental stress, which we will come back to time and time again in the story of our urbanisation; and when you put together the evidence for shrinking villagers with the other skeletal clues, the story of the Neolithic transition sounds like a pretty rough ride.

And it's a rough ride trying to understand all of this, especially squashed in the back of a Toyota making its knuckle-whitening way from the Dead Sea coast upwards some 1,200 metres (4,000 feet)* to the Jordanian plateau. After our encounter with the semi-subterranean roundhouses of Wadi Faynan 17, we headed over the edge of the Jordanian plateau and towards some slightly later sites, where Bill had promised some excitingly square houses – a later adaptation in the early days of settling down. Around halfway through the climb, however, one of the four-by-fours managed to finally rattle a battery free of its (in retrospect, poorly designed) plastic casing, causing the whole convoy to come to a shuddering halt on a hairpin bend, deep into the winding, rocky, Martian landscape of the wadi. The clown-car scenario repeated itself, as each of the trucks disgorged its quota of academics, all of whom took a deep interest in the stress mechanics of engine plastics but professed an embarrassing ignorance of how to deal with a truck battery no longer at one with its truck. But the wadis are never as empty as you would think – no sooner had people started talking wistfully of lunches then a little white pickup came roaring up the valley. Its driver was an unbelievably resourceful teenager, son of one of Bill's acquaintances from the Bedouin village. He popped out of the pickup (leaving a younger brother in the cab) and began an animated discussion with those members of the party who spoke Arabic. This led to the rather surreal tableau of a perhaps 13-year-old boy explaining to a bunch of middle-aged professors how to tie a battery up with string, which our pint-sized saviour handily accomplished. A quick phone call to his father to arrange for

* Yes, the Dead Sea is below sea level: 420 metres (1,400 feet) below; only one of the many things wrong with that particular body of water.

a new battery at the top of the hill and we were off up the valley again, while he reversed backwards down one of the most terrifying ridges I have ever been driven along. We just about caught sight of his father in the horrendous dust cloud he raised, a tiny speck in a black *galabeya*, standing on the valley floor and surrounded by his goats, waving his mobile phone in the air in greeting and, to all appearances, laughing uncontrollably in our general direction.

Soon enough we crested over the top of the plateau, ending up in the lonely and wretched semi-abandoned stretch of tourist-trap restaurants outside the entrance to Petra.* Cold glasses of lemonade with shredded mint (a strong contender for the title of world's best non-alcoholic beverage) were handed around, and the group began working its way through falafel, pizza and other traditional foods as we waited for the battery to be replaced. This was as informal a place as any for discussion of those big questions I had contemplated during the journey, so I spent the otherwise pleasant meal demanding answers from this gathered group of experts based on the myriad of sites they'd worked on. If we'd just seen the early experiments in settled living – and there's such a tranche of bioarchaeological evidence suggesting that settling down wasn't all that great health-wise – well … how could we make sense of it all?

The answer, for me, is to go back and look at those resources. How did sedentary communities manage their resources, and what effect did the major new resource of farmed food have on how they did this? Also, what about domesticated animals – meals on the hoof, so to speak?† For bioarchaeologists, the big question comes down to biology, and humans who as biological organisms either eat well or don't, grow well or don't, but either way leave telltale traces in their bones and teeth for me to find, thousands of years

* The Syrian situation had already deteriorated a great deal, and tourism was essentially flatlining; also, it was January and *snowing* in places, but that is an entirely different driving-related disaster story.

† Lots of people had ordered the *köfte* (meatballs); it was on my mind.

later. Even within that group of experts, who between them have probably dug up more evidence on the early Neolithic than any other conglomeration of humans alive, there were no easy answers. What drove the success of settled groups? How successful were they really, if they shrank in height and didn't live as long? What we do know is that the answer is more complex than just counting the bodies and measuring them up. We've got to delve even deeper into these Neolithic lives (and deaths) to understand the pretty radical trend of staying put.

CHAPTER TWO

Feed Me (Seymour)

We are very fortunate in this day and age to be living in a time in which mass communication has made it possible to share the scientific breakthroughs of our species, giving people a better grasp of the underlying workings of our universe and ourselves. We are slightly disadvantaged, however, by our limited attention span, which seems to want nothing more than to take the whole sum of human knowledge and distil it into an internet listicle of things that cause/cure cancer* or, more importantly, will make you fat/thin. This last obsession, with slimming to an 'ideal' weight, is an extraordinary situation to be in for our species, considering the first couple hundred thousand years of our existence (and our ancestors' before that, *ad infinitum*) was mostly spent trying very, very hard not to starve to death. It's only in the last century or two that obesity has become a mainstream option; previously restricted to those elites around the globe who could command the greatest resources (and food), obesity simply never really registered as a problem in the past. The case is rather different today, as the availability of cheap calories expands rapidly and our waistlines follow. The obesity epidemic is blamed on a hundred different things, depending on some arcane formula known only to news editors, and each new scientific discovery is heralded in the media as definitive proof that some particular item (a food, the internet, anything enjoyable) is solely responsible for the shape of the dent in your office chair.

Given the flurry of interest that stirs up whenever a putative cure or cause for one of modern humanity's most costly epidemics pops up in the news, it's not surprising that solutions to our obesity crisis with a certain 'back to basics' theme have

* Looking at you, *Daily Mail*.

flourished. Humans are masters of analogy and are very fond of linking small-scale, easily comprehensible narratives to more complex, abstract phenomena, and never mind the details. One of the most overused of these logical shortcuts is the one linking '*x* was better in the past' to 'we must do everything as we did in the past'.* Obesity is a very real problem, and the desire to do something about it, especially on a personal level, is easily understandable. It's not terribly surprising that, over the last few years, there has been increasing clamour about 'reverting' to a diet from a time when we were not all one bacon-flavoured cupcake away from type 2 diabetes. Cue the rise of the Paleo Diet™ and its many side branches, imitators and modifiers. These generally advocate a return to a putative 'pristine' hunter-gather state of health through the adoption of an allegedly authentic hunter-gatherer diet.†

At the time of writing I am utilising our capacity for mass communication to look online at the wrapper for something called the PaleoDiet Bar. It promises me optimal nutrition for a hunter-gatherer lifestyle. It is gluten-free, grain-free, soy-free, dairy-free, preservative-free, and fibre- and protein-rich. The presence of joy is not mentioned, but I think one could sensibly infer its absence from the litany above. I cannot actually get hold of one of these bars, as I'm writing this chapter from the middle of a recently resurgent guerrilla war in the southeast of Turkey, and on top of the roadblocks, the shootings and the heat, there don't seem to be any nutritional supplement/health food emporia this close to the border. However, the question remains: why does this product exist? Why is there a pre-packaged, plastic-wrapped, mass-produced snack endorsed by 'the world's leading expert and founder of the Paleolithic movement'‡?

---

* *e.g.* any sentence beginning: 'In my day …'

† See the excellent *Paleofantasy* by Marlene Zuk for a much broader discussion.

‡ According to Loren Cordain's website. It's unclear what the endorser is the world's leading expert on, but one suspects it may not be palaeoarchaeology.

Aside from the incredible unlikelihood of any human alive today having 'founded' the stone age – at last check, that would have been around 3 million years ago – it's a bit unfair to pick on this one particular instance of what I would like to start referring to as Neolithic denial. The 'palaeo' movement is indeed a popular one, and its starting point is that the development of agriculture has been one long, devastating mistake for our species. The key argument is that eating cultivated foods (ones that require human interaction, at least to some extent, to grow to desired numbers or amounts) is something we are not *evolutionarily* designed for. The cultivars we came up with – wheat, rice, legumes – are something our bodies are not able to digest, and many of our modern health problems are due to the toxic nature of our heavily gluten-based diet. Awareness of gluten sensitivity or gluten intolerance has skyrocketed in Western, affluent contexts in recent years, fuelling understandable interest. So the only way to solve our inexplicable constant health crises* is to revert to a diet that our evolutionary history has prepared us for. According to palaeo-diet experts, this includes meat, fish, vegetables, fruit and not much else. But is it true? And if so, what did hunter-gatherers really eat? The invention of farming – is it really making us all sick?

The study of human health and human diets is a great preoccupation of bioarchaeologists, myself included. What we eat, and how much of it, has a big impact on our bodies – from growth and development to death and disease. In the long timeline of human history, there are only a few points at which we can really identify a major change in the way we ate. The shift with the biggest effect on how and where we live our lives for the last 10 millennia has been the development of farming: the domestication of plants and the development of all of the very labour-intensive practices that go into making a living from the soil. For hundreds of thousands of years humans subsisted on more or less 'found' food. Of course, some of this

* Everyone has diabetes! Everyone has asthma! Everyone has allergies! Everyone has coeliac disease!

finding requires quite specialised skills and knowledge, and the archaeological record shows that humans have eaten a quite varied diet in the past, depending on location, environmental conditions and factors such as prey availability. It's simply not possible to declare there is one true pre-farming diet. One of our species' greatest tools for survival is our adaptability; we are omnivorous, and clever with it.

Take for instance the several groups who live by hunting and gathering in and around the Kalahari Desert in Southern Africa.* Famous for their traditional method of persistence hunting, where they track down ruminants at speed for days until the animals just give up and stop running from exhaustion, this long-range hunting technique has been explained as the evolutionary outcome of our unique combination of sweating and bipedalism. This is the model of human physical achievement that many people (and all stock photo editors) have in mind when picturing the 'hunter-gatherer' lifestyle. But gazelles are not the only food; when opportunity arises for these hunters, just about any animal will do, up to and including the slower but much pricklier rodent-of-unusual-size found in the region, the porcupine. Modern ethnographic research with the Hadza of West Central Africa has shown that particular groups of hunter-gatherers actually get about 70 per cent of their total calories from non-meat sources. Anthropologists Frank Marlowe, Collette Berbesque and colleagues have made a detailed study of the food in the Hadza camp, noting, weighing and watching what's for dinner. They found that meat accounts for only about 32 per cent of the total calorie intake, with the rest being made up of a combination of baobab (14 per cent), tubers (19 per cent) and berries (20 per cent). Surprisingly, the last 15 per cent of calories consumed by the Hadza comes from honey, which requires considerable effort, danger and

* Many of these groups have been referred to collectively as the San (lit. 'foragers', but with pejorative connotations) or 'bushmen' of Southern Africa, but there is a great deal of linguistic and cultural variety in this reductive grouping.

discomfort to acquire; nonetheless, it is a substantial (and much favoured) part of their diet.

Archaeological finds indicate that our ancestors ate a huge variety of foods.* There are bones with cut marks, boiled bones, burnt bones, charred seeds and even tiny bits of plants stuck in the plaque on ancient teeth that help us reconstruct our ancient diet. It's actually relatively recently that archaeological recovery has begun to focus on the microscopic evidence of past diets. This might go some way towards explaining why the emphasis in the early days of archaeology seems to have been solely on animal consumption, as the evidence of meat eating is largely identifiable by the naked eye, and the bones themselves big enough to be collected by archaeologists. The meat-eating caveman trope that 'palaeo-diet' proponents hark back to is, however, rather outdated. In the 1960s, archaeobotany (the study of plants that humans ate and used in the past) really came into its own. Though it might seem unlikely, some organic material does survive down through the millennia. Plant remains can be carbonised into charcoal, seed pods can be dried up and scattered through archaeological soils, and individual grains of starch and pollen can be retrieved from very carefully recovered bits of dirt and even dental plaque. These microscopic finds are the result of an excavation technique involving a flotation tank, where water is used to allow heavier (non-organic) material to separate from lighter (organic) bits, and even more recent techniques such as the analysis of dental calculus (plaque) – and the results have dramatically affected what we know about the human past, as archaeologists themselves have readily admitted:

> *The reader will note that our preliminary report on the 1961 season states confidently that 'plant remains were scarce at Ali*

* My colleague and former office mate Laura Buck spent several months delighting me with ethnographic evidence for the consumption of reindeer stomach, but inexplicably rejected my suggested article title of 'Reindeer tummies and the Neanderthals who loved them'.

> *Kosh'. Nothing could be farther from the truth. The mound is filled with seeds from top to bottom; all that was 'scarce' in 1961 was our ability to find them, and when we added the 'flotation' technique in 1963 we recovered a stratified series of samples totalling 40,000 seeds.**

Prior to the development of these advanced techniques for identifying the micro-traces of plant use in the past, archaeologists reconstructed our relationship with plants through archaeological finds such as storage jars and bins, and finds of tools used for harvesting or threshing.

In the first part of the twentieth century, V. Gordon Childe began to pull together the scattered pieces of evidence for a new type of plant use from archaeological digs all across the Near East and Europe. He theorised, building on previous work and his own encyclopaedic study of artefacts from the ancient Near East and Europe, that around 10,000 years ago there had been a major revolution in human societies. Hunting and gathering had given way to a new Stone Age invention: farming. Childe was a lifelong Marxist, an Australian émigré to England, director of the prestigious Institute of Archaeology at University College London, and perhaps the most influential figure in archaeology in developing grand theories on the how's and why's of the human past. His political views strongly influenced his conception of this major shift in human activity in revolutionary terms. This book in fact owes him a considerable debt, as two of his concepts are reiterated here:† the idea of the Neolithic Revolution, and the idea of the Urban Revolution. We will revisit the Urban Revolution in later chapters, but it's in the Neolithic Revolution that we find the very first examples of a way of living – sedentary, dependent on agriculture – that

* This quote and more information on the development of archaeobotany can be found on Dorian Fuller's website: https://sites.google.com/site/archaeobotany/

† Loosely, with a great deal of revision and far less revolution.

some would argue our body of evidence suggests we were better off without.

The Neolithic Revolution, as described by Childe, is not a revolution in the dramatic, anthem-singing mode of eighteen-century France or twentieth-century Russia. His revolution is the progressive outcome of friction between different aspects of human life, rubbing up against each other until they wear each other into entirely new shapes.* For Childe, there is one clear hallmark of revolution: demography. He sees a 'bend' in the line on the population chart as a sure sign that society has changed – that the societal gears have slipped and fallen into new positions. He is not the only one. Subsequent researchers have begun to piece together the skeletal evidence for population booms and busts, not just in Childe's area of interest (the ancient Near East, especially Mesopotamia), but in many different locations where settled life and cultivation have developed. This includes the very important work of Bocquet-Appel who identified a Neolithic Demographic Transition alongside the development of agriculture: population booms accompanied by corresponding rises in illness and mortality. These two basic concepts, of a revolution and the evidence left behind in bones, are at the heart of this chapter. The question we must answer is whether the major changes of the Neolithic carry a similar body count to the more guillotine-happy upheavals – whether the development of settled life and control of crops and animals are the critical first step on a slippery slope towards the second of Childe's revolutions, urban living, and thence on to a rising tide of disease and death. But first we need to take a look at what the Neolithic Revolution really meant.

---

* This is perhaps the most simplistic description of Marxist theory in the history of archaeology; for those with the stomach for more on Childe's Marxism and how it has influenced archaeology, I highly recommend Randall McGuire's 2006 article 'Marx, Childe, and Trigger' in *The Works of Bruce G. Trigger: Considering the Contexts of His Influences*.

For Childe, the Neolithic Revolution was the point at which humans became masters of their own food supply. But this is an insufficient description of the changes in human lives first occurring around 12,000 years ago in the ancient Near East. After all, humans have always presumably been in control of their food supply to some extent, or we simply wouldn't be here at all. Many societies that do not extensively farm still maintain small gardens or localised patches of tended wild plants: yams, tubers, rice, and maize, among others. Plant-based foods are a critical part of the human diet, and as with adherents to exclusively animal-protein-based diets today, people in the past would have been at risk of protein poisoning, or what has been called 'rabbit starvation': without a sufficient mix of nutrients, a diet consisting of meat (particularly lean meat) alone can cause debilitating symptoms in as little as three days, and death within weeks.

Plant-based foods vary among the many ecological niches humans have occupied, but there are several lines of evidence that show the long antiquity of plant eating in the *Homo* lineage, including the advances made by microscopic techniques. Many plants create tiny silicate structures called phytoliths; these come in different shapes and sizes, and because of their mineral durability can be traced in archaeological remains. Phytoliths from the Neanderthal occupation of Amud Cave at the margin of the Jordan Rift Valley show that the edible seed heads of grassy plants were part of the human story as early as 50,000 to 70,000 years ago. Fragments of plant starches found calcified in Neanderthal dental plaque have shown that not only were they eating grassy plants, they were eating cooked starchy plants too. These techniques have given us a wealth of archaeobotanical evidence for the exploitation of plant foods by modern humans as well. But at some point, our casual relationship with wild types of grasses and tubers morphed into a much more complicated, interdependent affair.

Wild types of the major staples of agriculture have been identified in several regions of the world, with different types of edible plants predominating in different ecozones.

In the Americas, the wild ancestor of maize (corn) has been identified as a particularly unwelcoming-looking spindly grass called teosinte, native to the Balsas River valley of Mexico. Teosinte seed heads have just a handful of armour-plated seeds, nowhere near as enticing as the fat kernels found on modern maize,* but archaeologists have found traces of them on grinding stones dating back nearly 9,000 years. Rice, the staple crop of half of the world's population, has gone through similar changes. DNA analysis by Bin Han and colleagues of different types of domesticated rices (long- and short-grain) suggests that the wild ancestor of both varieties comes from the Pearl River valley in China, and archaeological evidence of rice is present in sites along the Yangtze River from around 9,000 years ago. However, like maize, the rice found in early archaeological contexts is not quite the same as the rice we know today: the seeds are of variable sizes, and earlier rice had seeds that would shatter easily – a good thing for a self-propagating plant, but less helpful for human consumers. Dorian Fuller, who led research at one of these early sites in the Yangtze River valley, has suggested that rice domestication was a slow process that occupied thousands of years, during which time the population of the area largely depended on other sources of food.† Cereals like wheat and barley have deviated from their wild ancestors in the mixed oak and pistachio ecozones of the ancient Near East with similar mutations: seeds that don't shatter to make them easier to harvest; changes in seasonality, distribution and the way the plants propagate to make them easier to cultivate; and changes in seed size.

In fact, almost everything we eat today has been genetically modified by human intervention. Potatoes, yams and other tubers have been domesticated independently in

---

* Though, according to exacting scientific research by Nobel Prize winner George W. Beadle, they can be made to pop.

† As Dorian Fuller said in his 2014 *Nature* article 'Domestication: The birth of rice': 'Nobody mentions the acorns.'

several areas, but there are clear distinctions between the petite purple Andean wild potato and the mealy behemoth to be found under baked beans and cheese in most English cafes. Carrots, as many people will be aware, started off purple; watermelons began pink and about the size of a grapefruit; grapefruits, limes, lemons and oranges can be just about any colour between yellow and green but are in fact man-made hybrids of a handful of green wild citrus types; and almost every green we eat – from broccoli to kale – is a type of mustard plant. There are incredibly few commonly consumed plant foods that haven't been bred into submission to human tastes. We have a long history of genetically modifying foods, a fact that is generally conveniently left out of the debate on the ethics of using genetic engineering (tampering at DNA level) to modify crops. Or, if you wish to take it from the view of the plants, humans have been extremely successfully domesticated as part of the dispersal and reproduction methods of several different plant taxa. Exactly how long we've been messing with our food, however, is a subject that's still yielding new evidence – making us rethink what we know about the connection between the major changes in how people lived during the Neolithic Revolution and the development of farming.

The best-studied examples of the development of agriculture come from Southwest Asia (the ancient Near East), East Asia and the Americas. It was the 'Fertile Crescent' of Southwest Asia that Childe identified as the source of the Neolithic Revolution, and wheat and barley domestication does seem to appear several thousand years earlier in the region than that of rice in Asia or maize in the Americas. It's also the region I know best, having spent some time working on the fringes of the Southwest Asian agricultural phenomenon, on Neolithic sites on the Central Anatolian plain. Central Anatolia today is dominated by agriculture; fields of grain roll endlessly across the high plateau, disappearing off into the horizon in a wash of golden-yellow stalks broken only by the occasional irrigation channel. The wild ancestors of the modern suite of crops used in Europe and Western Asia can be found from Anatolia to the Southern Levant: einkorn, emmer wheat and barley, alongside other staple plant foods such as peas, chickpeas and lentils. That is, they can mostly be identified in these regions; the combination of evidence from modern crop DNA sequencing and the finds from archaeological sites paints a very muddy picture of the domestication process. Archaeobotanists are still searching for the wild ancestors of several modern food crops, like the broad bean, and results from DNA sequencing could be explained by a host of scenarios where wild and domestic versions of the same crop from different regions were repeatedly mingled over a period of thousands of years.

In 2015 a team of researchers working at the 23,000-year-old Levantine cave site of Ohalo II on the shores of the Sea of Galilee published new findings on the plant remains recovered from their site. These researchers didn't just look for the wild ancestors of our modern domesticates on site, they also identified the weeds that go with all of our tastier plants. By looking at the grouping of weeds and at infinitesimal scars on the seed heads, the researchers concluded that the residents had engaged in a sort of 'trial' farming, more than 11 millennia before Childe (and more than a few others) saw evidence of agriculture. The mix of weeds and edible grasses, the presence of sickles and the slightly domesticated shape of some of the seeds all suggest that the camp at Ohalo was an early experiment at crop cultivation – but one that ultimately ended up in failure. Like the Natufian experiment in settling down into one place, it seems that many of our initial attempts at revolution sort of fizzled out after a few thousand years or so. The 'Neolithic Revolution' seems to be more a 'Terminal Pleistocene Experiment', with a very gradually built foundation of new technology and lifestyles occasionally just razed to the ground while everyone goes back to hunting and gathering for a few thousand years.

In the 1970s, while researching the early agricultural group known as the Mound Builders,* physical anthropologists stumbled on a problem. Agriculture, most archaeologists reckoned, was a critical step on the progressive path towards civilisation. The sheer number of sites that pop up all over the world with evidence of agriculture clearly indicate that the human population had boomed with the advent of farming. For most, there was a clear trajectory from the 'revolution' of the Neolithic to the development of cities and all the technological advances our increasing numbers, in such condensed conditions, could come up with. With a rather uncritical concept of progress as 'a good thing' in much of archaeology, physical anthropologists suspected that they would find among the Dickson Mounds burials in Illinois, which included remains from both before and after the development of agriculture in the region, evidence of the benefits of this progress. What they actually found was a different story: the population grew, but at a price. Life expectancy was shorter in the farmers at Dickson Mounds, and life in general more risky. Childhood health and adult survival seemed to nosedive; how could this possibly be explained as progress? If the result of the Agricultural Revolution was a slide in living standards, why did it happen again and again, all over the world? How good was our new agriculture-based lifestyle, if we couldn't stick with it until 12,000 years ago? And how modified was it, really, from what humans did before? Here we come to the actual evidence for the effects of the Neolithic Revolution, taken from bones and teeth.

Researchers have been tracking human progress in the transition to agriculture from bioarchaeological markers for some time. A very early insight into the physical effects of agriculture came from the 'eloquent' bones of Abu Hureyra, an archaeological site in Syria that showed signs of human

* So called because they liberally littered the centre and southeast of the modern US with impressive earthworks from around 5,000 years ago right up until the contact period of the 1500s.

occupation from the hunting and gathering Natufian period through to early agricultural experimentation. Theya Molleson, a pioneering physical anthropologist from the Natural History Museum in London, researched the remains of those early agriculturalists. Molleson described a suite of changes to the bodies of the Abu Hureyrans, a build-up of the bone in locations where muscles might take the strain from repetitive heavy loads. For instance, she identified changes in the neck vertebrae where the weight from a heavy load carried on the head would fall; collapsed vertebrae right at the arch of the spine; and, of all things, an uncommon number of cases of arthritis of the big toe. Alongside these peculiar pathologies, she found considerable evidence of muscle use on the arms and legs; the parts of the bone where the big muscles attach were very built-up. Heavy muscle use encourages the bits of bone where the muscles actually anchor themselves onto the skeleton to expand their surface area in order to attach more tissue: on a skeleton, this might be identified as extra bone formation at the locations where muscles insert. Rightfully rejecting an early initial theory of Neolithic ballerinas, Molleson identified a pattern of wear and tear very specific to the act of grinding grain on a stone quern. Holding this original kneeling 'plank' position while grinding grain had devastating effects on bodies, particularly those of the women of Abu Hureyra. While later studies have questioned the extent to which any habitual activity really alters the structure of the skeleton (something we'll discuss in detail in Chapter 13), it is clear that Molleson had identified a characteristic suite of actions that had real (and lasting) consequences for Neolithic people.

There is strong evidence for a downside to the Neolithic lifestyle. Early anthropologists noted that many population groups in the past had remarkably straight, healthy teeth. There were far fewer criss-crossed incisors, straggling canines or sideways-sprouting wisdom teeth.* Searching for

---

* And, presumably, far fewer dentists.

an explanation, many researchers identified changes in the way we use our teeth as the cause of our recent dental distress. One of the things researchers have noted is that the very hard enamel surface of teeth carries a tiny legacy of microscopic scratches and pits from the many things we chew. Peter Ungar pioneered new methods of studying dental microwear using high-resolution images of teeth blown up to a size where these scratches can be counted and described, and has suggested that by distinguishing between long thin lines, deep scratches and pits, different types of diets* can be identified. A switch between a diet of hard nuts and seeds (leaving lots of microscopic pits) and a diet more focused on softer fibrous plants like tubers (leaving more scratches) should change the patterns seen in the enamel under the microscope. This is in fact what has happened: there are reports, for instance, of an increase in pitting that are attributed to an uptick in the amount of hard nuts and seeds consumed in the long Neolithic transition of the Late Archaic/Woodland people of North America. Patrick Mahoney, looking back at the ancient Near East, sees the same phenomenon, but adds a few notes of caution – different foods probably require different amounts of chewing, so he tested the idea that soft grassy seed heads wouldn't change too much with the shift to a Neolithic diet among the Natufian people, but what would change is the pitting on their teeth. The change in pitting could be attributed to the archaic trend (recently revived) for stone-ground grains. The same stone-grinding techniques that were building up the muscles of the women of Abu Hureyra were leaving microscopic traces somewhere else: the near-invisible bits of stone grit you get from bashing rocks against each other went straight into the food, which went straight onto the teeth.†

---

* And even tool use – in a world before the invention of the table clamp, teeth were frequently pressed into service as a third hand.

† And still does; modern stone-ground flour is also likely to contain grit from the grinding process.

Many foraged foods – for example, the baobab mentioned above – are fairly tough, fibrous options and require a considerable amount of chewing in order to obtain any nutrition out of them. The soft and mushy cooked carbs that came to predominate our diet with the advent of agriculture, it was theorised, were rather less taxing on our jaws, even if the stone we ground them with did leave craters in our teeth. This created a sort of Pot Noodle effect: soft, slurpable foods means less work for our jaws, which means that less muscle needs to attach to them, which means that they don't need to be as big to support the muscle; and if the jaw doesn't need to be big, then perhaps the face doesn't either. And shrinking faces is exactly what we get, according to observations from different studies by prominent physical anthropologists such as Simon Hillson, Clark Spencer Larsen and C. Loring Brace. Shrinking faces would be neither here nor there in terms of positive or negative effects of agriculture, but the problem is that our faces have our teeth in them. Teeth are under quite strong genetic control, responding far less plastically than the rest of the skeleton to changes in use and environment. If the teeth stay the same size in smaller jaws, or even just shrink more slowly, you're going to get overcrowding, and the malocclusion (teeth pointing every which way) observed in many remains from the cusp of the Neolithic. While there's probably a strong component of genetic luck to how well our teeth fit in our jaws, those of us who have had to have our painful wisdom teeth yanked out might want to send the dental bills to those early farmers who kick-started our easy-chewing diet.

It's not just the size of our jaws and teeth that have changed in the last 12,000 years. Our overall dental health has taken a pretty severe beating. Caries (or cavities) are the holes that enterprising bacteria excavate into our teeth given the right environment. These holes can expose the nerve endings at the heart of our teeth, leading to sensitivity (to heat, to cold, to contact) and occasionally rather excruciating pain. While our teeth do have a built-in defence mechanism, building up bulwarks to try to protect the nerve when the hard outer enamel is eaten away by the lactic acid emitted by well-fed bacteria, caries can work faster and lead to considerable destruction, even the loss of the whole tooth. The determining factors in this whole painful process are the combination of oral bacteria (something you are likely to more or less inherit*) and the foods you feed them every time you put something in your mouth. Caries bacteria are pH-sensitive, thriving in a base environment and less active in an acid one. When you eat, the pH balance of your mouth changes according to the type of food you need to digest. Eating starchy, carby, sugary foods is the best way to encourage caries bacteria – the sugars generated by eating these foods depress the mouth's pH balance for a longer time, allowing more opportunity for caries to develop. So when we see teeth from the past riddled with holes, there is good reason to suspect that a starchy, sugary diet might be at work – and when it comes to the development of agriculture, we start to see an epidemic of rotting teeth.

Of course, we have good evidence that humans have been getting lots of their calories from carbohydrates for a long time. In Morocco, mobile groups from the Iberomaurusian tool-using culture (not farmers at all) have fairly wretched teeth. Caries seems to have been a big problem for these hunter-gatherers. Researchers identified the likely cause as acorns – a good source of nutrition, but a poor choice from the point of view of non-rotting teeth. Another hiccup in the easy assumption that 'caries equals farmers' is the variation in caries between different groups who are eating similar foods. As DNA analysis techniques continue to improve, we might find that different strains of caries-causing bacteria are more or less virulent, and that the kind of oral bacteria you inherit might have a considerable effect on your teeth's survival. Natural fluoridation of local water sources also plays a role in protecting teeth from caries. While rotting, misaligned teeth are not necessarily the smoking gun of agricultural innovation,

* Caries bacteria are usually transferred directly to an infant – pre-chewed food and maternal affection being prime culprits.

the numbers do however suggest that they start to become a real problem for our species round about the time we develop agriculture.

So, how risky was the Neolithic Revolution? It might mess up your teeth, but is that enough to kill you? Given the uptick in births and the fertility 'revolution' that researchers have identified during the transition to agriculture, how do we understand all the factors that kept our numbers down? This is a question that colleagues and I have tried to address by looking at the life history information that is locked into the hard enamel of human teeth. Teeth are a wonderful resource. Made of about 98 per cent mineral, they are robust and durable in most archaeological soils, and may survive thousands of years beyond the more fragile bones. Teeth begin to form before birth, and never remodel,* unlike bone, so they carry the chemical and physical signature of the time when they were growing with them forever. The enterprising dental anthropologist has a range of techniques available to try to reconstruct these signals, giving us the rather unique opportunity to look at human lives in the past, rather than human deaths.

If you are at all familiar with the concept of tree rings, you can imagine a similar scenario at work in your teeth. A tree grows in successive layers, each bounded by a ring as the tree trunk expands year-on-year. In good years, the tree grows quite a bit, and the ring formed is larger; in bad, it grows less, and the ring is smaller. While the analogy is loose,† your teeth also form in similar successive layers.‡ The layers

---

* Anyone who has ever chipped a tooth will be dramatically aware of this fact.

† Should my old PhD supervisors ever come across this section, I fully expect reports of eminent physical anthropologists spontaneously combusting to rapidly follow.

‡ And in multiple rows, like a shark; baby or milk teeth start growing before birth and the wisdom tooth finishes around age 15. The in-between stage, where both sets are present to some extent, is truly terrifying in X-ray.

respond to a sort of innate rhythm, an internal timer that runs at about a cycle a week, and where growth stops (and restarts), it leaves a little ring around the tooth. Anyone over 18 will probably have brushed away most of these little rings, but occasionally they might still be visible with the help of a strong light and a good mirror. Being able to count the rings on teeth is of course useful in the same way that counting rings on trees can be – it tells you how long the item in question has been forming. But more than that, it's when the rings go missing that we start to get real insight into what was happening when your teeth were growing. Where there are gaps in the normal pattern of the rings – basically, depressed grooves or lines on your teeth – it's a sign that normal growth shut down, usually due to illness, possibly disease, or even malnutrition. These grooves on teeth are called enamel hypoplasia, and they can tell bioarchaeologists whether the child who was growing the teeth was healthy – or rather, when it was not.

Researchers have been aware of the connection between lines on teeth and childhood health for some time. It took the people of Dickson Mounds, however, to really emphasise its importance in tracking how human populations deal with changing circumstances. This 1970s study was one of the first to illustrate that evidence of childhood illness and malnutrition locked into the teeth could be compared between farmers and non-farmers; and when the results were in, the farmers had far more enamel hypoplasia than the non-farmers. There is a general agreement that the transition to agriculture leads to an increase in the number of lines on teeth – this is observed not only in the Near East, but almost everywhere that farming kicked off. This is a critical point – if we have conflicting evidence for how taxing the transition to agriculture was, can we really make such sweeping universal statements? It's what archaeologists do,* but we can look to different areas of the world and different *types* of Neolithics to see that the

* Especially ones with the temerity to write books for a popular audience.

observations that hold true on the East Coast of the US don't necessarily match those elsewhere, like Thailand. As technology improved, it has been possible for bioarchaeologists to get ever more detailed information about when and how children were sick in the past; so, armed with some dental kit and a blind faith in my ability to drive across Anatolia, I set off in 2012 to have a (much) closer look at one particular site: Aşıklı Höyük.

Aşıklı Höyük is a great mound of earth built up on the sides of the Melendiz River on the fringes of Turkey's mountainous Cappadocia region. It's closest to the modern-day city of Aksaray, but still a lengthy* drive from just about anywhere. Excavations on the mound began in 1992 under the direction of Ufuk Esin, a pioneering Turkish archaeologist, and continue today under her former student, Mihriban Özbaşaran of Istanbul University. I had met Mihriban and her team while we were both working at the UNESCO World Heritage site of Çatalhöyük in 2008. Çatalhöyük is another mound site on the Anatolian Plateau, and rather better known – decades of excavation there have unearthed a warren of mud-brick houses decorated with painted plaster and cattle skulls and with underfloor burials.† Dating to nearly 9,000 years ago, Çatalhöyük is billed as one of the earliest 'cities' in the world; at its peak, perhaps 10,000 people lived and farmed together. A crack team of physical anthropologists from all over the world gathered at the site every field season to investigate the remains of these early inhabitants, and I was fascinated by the opportunity to learn

*Not to mention slow. It once took me nearly 30 minutes to traverse the small village of Kızılkaya near the site, thanks to successive traffic jams caused by cows, women herding cows, geese, women leading donkeys to follow the cows, chickens, women following the women leading the donkeys following the cows, dogs, and, just in front of the gate, the world's least motivated tortoise.

†While this might sound like the description of a particular kind of rural drinking establishment, the decor scheme is actually slightly more apocalyptic – there is an erupting volcano mural, for instance.

more about these lives on the edge of the Neolithic Revolution. I was therefore understandably excited when I learned that the Istanbul team was digging another settlement site in Anatolia – and this one was even earlier.

Despite an inauspicious introduction,* Mihriban and her assistant director Güneş Duru graciously invited me to visit Aşıklı in the summer season of 2012. Aşıklı was turning out to be an enormously important site. Radiocarbon dates showed that the first phase of settlement on the mound was around 10,500 years ago – nearly 1,000 years before the hubbub at Çatalhöyük. What's more, the site covered almost a millennium of occupation, from early roundhouses that were just a tad more permanent than the seasonal encampments of contemporary hunter-gatherer sites, to a full-blown mud-brick village with wide public spaces and the scythes and storage jars that are the hallmark of an agricultural lifestyle. Of course, nothing would do but for me to come look at the human remains. Here was a chance to look the Neolithic Revolution straight in the face – though in my case, it was the teeth I was really interested in. With the support of the British Institute at Ankara, and Yılmaz and Dilek Erdal at Hacettepe University, I took dental impressions from the dead then carefully carried the impressions back to a basement lair at the Institute of Archaeology at University College London. There I spent a rather tedious amount of time in a sunless room carefully counting lines on teeth.

Fortunately, this was time well spent. The Aşıklı teeth showed a pattern, very faintly, of interruptions to the normal pattern of growth lines. The teeth that came from individuals from the later period of the site showed big grooves where growth had been interrupted, occurring about every two years starting around the age of two. One lone individual, however, had a slightly different signal – there was more evidence of growth disruption around the age of three. This was the one skeleton from the earliest phase identified at Aşıklı – the phase of roundhouses and ambiguous evidence

* I may or may not have accidentally destroyed a 9,000-year-old wall.

for dependence on agriculture or domesticated animals. While one skeleton does not a conclusion make, it's tantalising to consider that the problems these children had growing up are linked to the same changes identified by Bocquet-Appel and colleagues: more babies, and more often. For many primates, the most dangerous point in childhood is when the mother turns her attention to the next infant; problems with food supply and illness can accompany a newly free-range child. In Aşıklı, it's possible that the timing of this rude interruption to childhood health related to the birth of new siblings. In an experimental time, without easy access to branded snack foods (gluten-free or otherwise), it might be that the fertility unleashed by settled life took its toll on the health of the children of the revolution. In the meantime, we must wait while more evidence is uncovered, and excavations at Aşıklı are ongoing.

So our universal ideas of the Neolithic may not be so universal after all. Physical anthropologist Dan Temple has spent *nearly** as much time as I have counting up lines on teeth. Working with Clark Spencer Larsen at Ohio State University on the skeletons and teeth of foragers and subsequent farmers who inhabited prehistoric Japan (the Jomon and Yayoi cultures), Temple has reported that the adoption of agriculture didn't lead to more lines on teeth or other markers of poor nutrition. The only negative sign he has found in the Japanese Neolithic transition is evidence of a slightly increased infectious disease burden. It is interesting that wet rice agriculture seems to have this effect not only in Japan, but in the well-documented Neolithic transition of Thailand.

As bioarchaeological techniques become more sophisticated, archaeologists will be able to tell more about the survival and fitness of the earliest farmers. Advances in aDNA analysis can also shed light on the success – on a population scale – of Neolithic ways of life. Sophisticated modelling techniques offer new insights as well: recent work by a group led by

---

* This is my book. He can say he's winning in *his* book.

Steven Shennan has tracked an enormous number of early farming sites in Europe, and their data shows a 'boom and bust' pattern of expansion of agriculture-dependent lifestyles. The idea of a steady march of progress as a straight line between the invention of farming and the height of modern civilisation* has been shown to be more of a St Vitus's dance – jittery, unpredictable and with a pretty high body count at the end.

---

* As originally conceptualised, this generally means the British Empire. Though occasionally, also the French. Absolutely no one has ever used this concept to describe a world with reality television in it.

CHAPTER THREE

What's New Pussycat?

Humans have manipulated thousands of years of animal evolution to make a tastier chicken, a milkier cow and a wolf you can let play with your children. Is it really very surprising that the changes we've made to the way we interact with animals have also impacted us? Our ever-closer relationship to animals has had unintended side effects, both good and bad. For instance, some researchers believe that humans would never have reached the numbers they have without the help of domesticated animals. On the other hand, there is hard evidence that living in close proximity to livestock has brought humans a host of trouble as well – diseases, bugs and parasites. Understanding how we have changed animals might come from identifying the leftover bones found in early human habitations, tracing the genetic legacy of our food and friends, or even mapping how animals have moved around the world. But what many of us may not have realised is that animals have left their mark on humans as well. Life with animals has gotten into our very bones.

In July of 2015 I found myself on the small, remote Greek island of Antikythera. It's an extraordinary place: approximately 20 square kilometres (8 square miles) of rock thrust up on the westernmost edge of the Aegean Sea, about halfway between the ever-so-slightly larger island of Kythera (avowed birthplace of Aphrodite) and the westernmost finger of Crete (avowed birthplace of Minotaurs). Bronze Age settlers from Minoan Crete colonised it, then apparently thought better of the enterprise. A group of Cretan pirates in the late fourth century BC set up a franchise on the island that was a bit too successful, raiding ships coming through the notorious Straits of Kythera. It's not clear from historical sources exactly what happened, but there are unused metal slingshot balls lying in the ruins of the fort with inscriptions reading something like 'from the

Phalasarnans', a group of related pirates from Western Crete, which, while witty, failed to save the island from another period of abandonment when the Roman general Metellus went around clearing the sea of pirates. Farmers came and went, as did colonial administrators; the latest inhabitants again trace their roots to Crete in the late 1700s. Throughout these thousands of years, island life had gone through constant shifts and phases: it's a bit more pirate-y at times and a bit more agrarian at other times, but all through the waves of abandonment and resettlement, one thing remained constant.

Goats.

It's the first time that I've been back to Antikythera since spending three field seasons walking over every inch of this scrubby, rocky landscape in the mid-2000s for an archaeological survey project. That experience was sufficiently strenuous that I've reduced it in memory to a haze of cuts and scrapes, local wine with a petrol sheen glossing the top, and the inevitable choice of goat or goat for dinner. Goats, actually, are responsible for the cuts and scrapes too. Not directly, despite their off-puttingly square pupils and the general aura of menace as they slowly surveil you from a ridgeline high above; but the blistering poison oak rashes, the thorns embedded in socks, skin and (impressively) leather paratrooper boots – these are all their omnivorous, insatiable fault. Goats, it is well known, will eat anything. What is less well known is that, given an island more or less to themselves for 5,000 years, goats will eat the soft, pleasant parts of Mediterranean vegetation to near extinction. What survives are the toughest, spiniest, downright orneriest of maquis scrubland plants: kermes oaks (poison!), salt-stunted pines and something we called 'born-dead' (probably *Sarcopoterium spinosum*, putative material for Jesus's crown of thorns). This dense thatch of scratchiness was excruciating to walk our straight survey lines through. As someone who doesn't eat a lot of meat, I have to admit that when the local taverna owner Myronas (who also doubles as the postmaster, harbourmaster and proprietor of the general store – it's a *very* small island) served up the nightly plate of boiled goat, it tasted a bit sweeter for the faint savour of revenge.

Antikythera is an island shaped by goats. The food, the industry,* even the landscape itself is heavily goat-based. There are several thousand goats on the island today, and the islanders divide them into rough geographic herds but no longer directly manage them. There are no milking goats left, only the freest of free-range animals, scrambling over vertiginous sea cliffs, napping in crumbling old houses and looming over the abandoned and overgrown fields from every ridgeline. The goats we ate roamed the empty dirt roads of the tavern keeper's old village, and Myronas would wander up the hillside in his flip-flops, dog at his side and shotgun over his arm, waiting for them to make their way towards an old fenced-in watering trough that he'd habitually open up at the same time each afternoon. Once there, he'd open the gates and settle down to wait. Eventually we'd hear a loud bang or two, and it would be goat for dinner the night after. Myronas can always count on goat as a safe bet for dinner, even when the boat that brings most of the island's food doesn't land because of the precarious anchorage of the north-facing harbour.† And they are *his* goats, in the same way that the carefully managed beehives in a nearby field are his sister's beehives. Antikytheran goats actually have some degree of local fame – they are judged to be tastier than the more

---

* When there was some – the island is nearly abandoned now with only about 40 or so hardy souls remaining year-round.

† Something that happens with alarming regularity – it's entirely possible to get stuck on Antikythera even in modern times. At the time of writing this chapter, there is a Dutch-Turkish couple that cannot get their sailboat out of the harbour until the wind changes, something that's unlikely to happen for at least another few days. In 2005 one intrepid project member, Ismini Papakirillou, made a rather dashing dawn escape by jumping into a speedboat of unknown origin heading for Crete.

carefully managed herds of other, more populous islands. They also look a bit different – shaggier of coat and ever-so-slightly bigger horns.

But, and this is the interesting bit – are they really 'wild' goats? Was Myronas, with his shotgun and sandals, hunting and gathering? We look at the scope of human history and our relationship with animals and tend to see a progression starting somewhere in red ochre paintings of wild aurochs and somehow winding up in the controversial practices of contemporary industrial meat production. But the story of domestication is not so linear. If we want to understand what animals have done to us, we have to trace back through our shared history to see how we first came to keep them close.

What is a domesticated animal? How do you spot one? Aside from the obvious tests (*e.g.* 'has it got four legs and has someone put it in knitwear/a handbag?'), there are some subtle changes that have been noticed across species. Darwin himself in *On the Origin of Species* picked up on some of these outward indications of domestication, insisting: 'Not a single domestic animal can be named which has not in some country drooping ears.'

While floppy ears may not be the defining characteristic of all domesticated animals,* there is a clear suite of physical changes that mark out many domestic animals from their wild ancestors.

These include the more ephemeral behavioural changes we associate with domestic animals, alongside structural physical differences. 'Tameness' as a trait is an interesting concept that has actually been explored in detail by the geneticist Dmitry Belyaev, who dedicated his entire research career to the experimental domestication of the silver fox at the Institute of Cytology and Genetics in Novosibirsk, Siberia. Researchers selectively bred foxes over generations and generations (50-plus, and the work is ongoing) for their willingness to engage with human handlers. After half a century of research, nearly all of these

* For instance, I have yet to observe floppy ears on a chicken.

selectively bred foxes are willing to socialise with humans, and the internet has video proof of their wagging tails and dog-like temperaments. The experiment called for selection on only one trait – behaviour. The less aggressive towards a human handler a fox was, the higher the animal's ranking, and the highest-ranking foxes were selected to breed a new generation. This was repeated in each new iteration of the breeding programme. After eight generations, the selection for 'tameness' had a noticeable effect on fox personalities. Kits were far more likely to show affection to and seek attention from humans, a reasonable outcome for selection of a behavioural trait. But their behaviour wasn't the only thing that changed. Foxes were born with pigment losses that gave them white blazes and face masks like a collie, with brindled coats and curling tails. Floppy ears, Darwin's harbinger of domestication, also appeared. The domestication process had not only changed the foxes' basic sociability towards humans, but had changed them physically as well. These physical changes weren't directed, but just unlooked-for side effects of humans favouring friendly foxes.

Since evidence of behaviour and floppy ears does not survive well in the archaeological record, it's the changes in bones and teeth that become key to spotting domesticates in the past. Zooarchaeology is dedicated to the identification of animals found alongside evidence of human activity, and much of this research includes picking up the subtle differences between the teeth and bones of thousands of past meals to explain what and how we ate. This kind of research tells of a long human history of eating not only the big-game animals we tend to imagine when the word 'hunting' is bandied about, but also shows us the delicate bones of a pigeon, the spine of a fish, and countless other small animals besides; prehistory is revealed as an opportunistic open buffet. In identifying these countless species, however, zooarchaeologists also get a chance to look at the subtle signals indicating that an animal has been domesticated. This can mean analysing the shape of the bones to look for variation from the wild type, or more complicated analyses that look at the age distribution of the animals on site, the isotopic evidence of the food they have eaten and other tangential clues as to human intervention.

Dogs provide a very helpful example to illustrate how the actual skeletons of domesticated animals can give them away. Even though we might easily recognise a dog when we see one walking down the street, they have been bred into such fantastic shapes that their skulls and bones look as though they are barely the same genus, let alone species.

Many of our fancy breeds of dogs have extreme – and deliberate – modifications to their skeletons. The shape of the jaw and skull, carefully selected for breeds like pugs and Pekingese, is diminished and distorted into a gross overbite, resulting in misaligned and even missing teeth. This can also result in breathing difficulties for the animal of varying degrees of severity, and concomitant 'cute' traits such as snoring. The normally longer-than-it-is-wide shape of a canine skull has been modified in several breeds to the opposite, a wider skull, which carries a host of consequences for breathing and, in the case of the Cavalier King Charles Spaniel, means bits of brain squishing out at the back of the skull because the fit is too tight.*

How does domestication happen? There are a variety of theories on the subject, and it may be that there is no one way that animals are modified into domesticates. The human habit of bringing animals home in live (rather than carcass) form is surprising both in its antiquity and ubiquity. There are several groups around the world who still subsist by hunting and gathering, and ethnographers have observed that in many of these cultures the animals are frequently brought home as 'pets' (though they tend not to fare particularly well). Pet-keeping as we understand it now – a

* A condition called syringomyelia, which leads to debilitation and death in the animal, and furthermore makes a strong argument for responsibility in dog breeding.

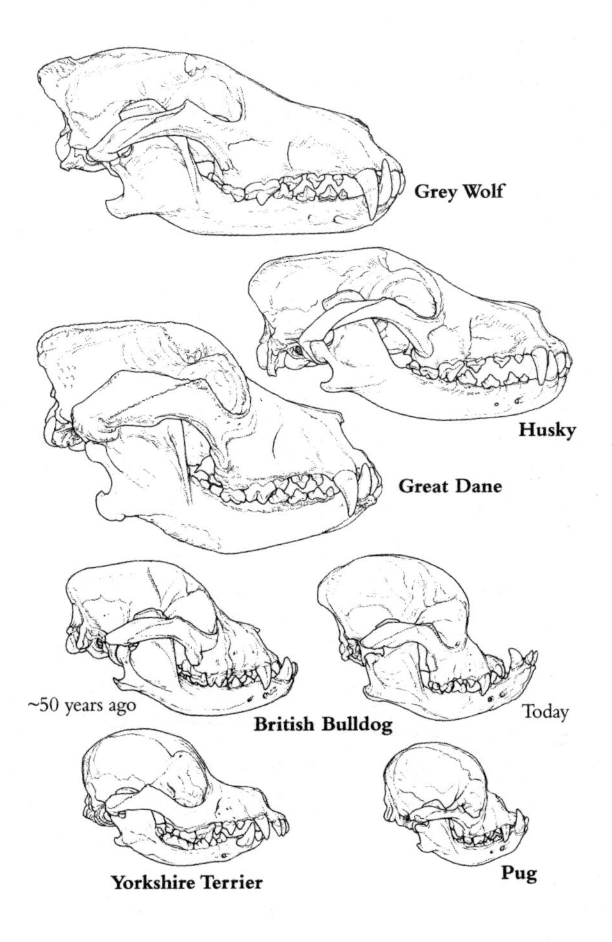

Figure 2 *Selection of canine skulls.*

form of companionship rather than a working relationship – was in the past in many societies restricted to elites,* such as Chaucer's Prioress, Madame Eglantine (who kept small hounds fed on meat and milk), or the famously

* History is absolutely full of fascinating pet stories from the elites of ancient Greece, Rome and other cultures; this includes the utterly unexpected upper-class Roman tradition of keeping turbots as pets, and occasionally decorating them with gold rings.

spaniel-loving court of England's King Charles II. So popular were his pet dogs that they were constantly being stolen, and special pleas for their return had to be made. Among the very earliest printed examples of a 'missing dog' ad is this impassioned plea on behalf of the king:

> *We must call upon you again for a Black Dog between a greyhound and a spaniel, no white about him, onely a streak on his brest, and his tayl a little bobbed. It is His Majesties own Dog, and doubtless was stoln, for the dog was not born nor bred in England, and would never forsake His master. Whoesoever findes him may acquaint any at Whitehal for the Dog was better known at Court, than those who stole him. Will they never leave robbing his Majesty! Must he not keep a Dog? This dog's place (though better than some imagine) is the only place which nobody offers to beg.*
>
> Printed in the 28 June 1660 Mercurius Publicus and quoted in Notes and Queries 7th S., vii. 26.

Dogs may have in fact been the very first animal domesticates. Wolf bones have been found close to human remains, and indeed to the remains of our earlier ancestors as well. This is actually not so surprising – wolves and humans can both occupy similar ecological niches and hunt similar types of animals, so their presence in the same place might be explained by both species exploiting access to game, or even wolves scavenging the remains of human hunts. However, from about 32,000 years ago, we have the evidence of a very different kind of wolf found in conjunction with humans. A fossil skull of a canid from Goyet Cave, Belgium, has a shorter muzzle, wider skull and reduced fang teeth compared to most ancient and modern wolves. This, however, is not the ancestor of the modern domestic dog. It's actually not even very closely related to the grey wolves from which most dogs are descended. The most recent common ancestor of the Belgian canine and all modern dogs is estimated to have lived nearly 80,000 years ago, a previously unknown sister-group to the European grey wolf. But other dog-like remains abound in the Palaeolithic, as well as traces of human–dog interaction, like the large canid that left its footprint next to that of a human child some 26,000 years ago in the brilliantly painted Grotte Chauvet-Pont d'Arc, one of the earliest sites of ancient art.

Using analyses of ancient dog DNA, we can start to unravel the disparate history of these dog-like and wolf-like skulls found in cave sites and even burials to see multiple episodes of dog domestication. The Belgian discovery may not have led to our modern breeds, but shows that the domestication process may have been at work from multiple centres in many different periods. Wolves may have been domesticated down a 'commensal' pathway, adapting to live in and around human ecological niches and eventually developing a sufficiently different set of behaviours in order to exploit our good graces, which ecologically (if not, in fact, genetically) cut them off from their wild ancestors. Like Belyaev's foxes, domestication could have rapidly changed the physiology of captive wolves into dogs.

Dogs are commensal animals, like the ubiquitous urban pigeon or the louse. They require a human habitat to survive. The house mouse is another such animal; unsurprisingly, it appears in the archaeological record in association with humans around the time we begin to settle down and keep all of our food in one place – a house. Cats are also commensal,* but for a variety of reasons have not picked up the suite of human-centred behaviours that dogs have. Evidence of cat domestication, such as it is, only goes back to about 9,500 years ago, with the burial of a cat and human together on Cyprus. It may be that human ecological niches were not of great interest to cats until we started attracting so many mice; and they have retained their independence by constant interbreeding with the wild type of cat. While it's readily apparent from a quick perusal of popular culture, internet memes and Instagram accounts that many of our

* Though they probably wouldn't admit it.

long-term commensal animals have considerable impact on the human psyche, it's difficult to argue that our long history of living with dogs has had a major effect on human health. The diseases we get from dogs are fairly rare, and despite the dog's reputation as a carrier of disease and filth in some corners of the world, adopting dogs into our lives doesn't seem to have killed us all off. Interestingly, despite their omnipresence in our modern, digital lives in the form of a million wittily captioned pictures, cats seem to have very little effect on human health outside of the threat of transmission of toxoplasmosis. Some researchers suspect that *Toxoplasma gondii*, which is normally hosted by cats, has a subtle behavioural effect as well, for instance transforming infected rats from cat-fearing to cat-loving (and thence to cat-food). Jaroslav Flegr, who researches the bacteria, even suspects that toxoplasmosis infection, like rabies in dogs, may have similar, subtle behavioural effects in humans – the so-called 'Crazy Cat Lady Syndrome'. But these diseases do not, on their own, take down large numbers of humans, nor do our dogs and cats necessarily contribute to our survival (Lassie excepted) in the same way as our other, more consumable domesticates. Rather, the animals that have made the biggest impact on our lives are, of course, the ones we eat. These are the prey animals that we have adapted to accompany us, the 'meals on hooves' that have contributed to our calorific wellbeing to an incredible extent.

At Aşıklı Höyük, mentioned in the last chapter, the early occupation of the site is littered with a wide array of diverse animals – big, small, feathered, furry, spiny or shelled – that demonstrate the incredibly broad spectrum of the true 'palaeo-diet'. However, by the later levels at Aşıklı (approximately 6,000 years ago), the animal remains are dominated by a single group: sheep and goats. Moreover, the bones of these animals tell us that the males were being consumed at younger ages while the females were left to breed – under the watchful eyes of the inhabitants. A number of foetal sheep and goat remains found at the site testify to the active management of this food source. There are no newborn aurochs or horses on site, but there are newborn goats, suggesting that over the course of Aşıklı's long history, the inhabitants began not just to exploit sheep and goats but to truly domesticate them. This goes beyond the animal management strategies of different hunting and gathering groups, who might for instance hunt only male animals in order to maintain population numbers. The presence of newborn animals within the settlement and the micro-chemical detection of animal poo deposits at Aşıklı give a clear picture of animals that are being 'kept' – stabled – in a deliberate management strategy that may have included controlling both feeding and breeding of sheep and goats.

It is remarkable that of the multitude of species preyed upon by humans, only a very select few have been domesticated.* A variety of reasons have been proposed for why, for instance, the zebra has never been successfully domesticated when the closely related horse and donkey have. Jared Diamond has suggested some innate difference in animal temperament, a quality of 'tameability' that is either present or absent; in the case of the zebra, he reports: 'Zebras are incurably vicious, have the bad habit of biting a handler and not letting go until the handler is dead, and thereby injure more zoo-keepers each year than do tigers.'

The actual answer may lie somewhere between this extreme position and the variety of ecological niches that animals and humans inhabit. For instance, dogs and humans share similar habitats, and domestication can be seen as a reasonable outcome of the considerable overlap of scavenging species with big-game hunters such as our Palaeolithic ancestors. Domestication of our prey animals such as sheep and goats, however, seems to have been motivated by the necessity of trying to manage the *where* of our meat supply when reliance on cultivation in a localised area began to take hold. The remains of animals eaten by the Natufian groups of 12,000 years ago indicate that

* Though this may be increasing; for instance, changes associated with domestication have been observed in the hamster, which might be expected from its popularity as a pet, and in the trout, which might not.

they were hunting and gathering, but buried under the mass of gazelle bones are the smaller bones of new groups of edibles: small mammals, birds and other more labour-intensive prey. Many have interpreted this as a sign that staying in one place put increasing pressure on local food resources, and a sort of *nouvelle cuisine* may have been required. Cattle and pigs were also domesticated in the Near East, as well as possibly Asia, by 10,000 years ago. The only exception to the pattern of sedentism and domestication is in fact the festive reindeer: reindeer herders in the very northern extremes of Europe are largely nomadic, and their managed herds exist alongside the wild type, leading some researchers to suggest that reindeer are in fact only semi-domesticated.

Exploiting animals for food has a long history in our species. So long, in fact, that there is considerable argument about which of our ancestors ate meat and when. The jaws and teeth of modern humans are adapted to be omnivorous – we have ripping, tearing teeth at the front and chunkier mashing and grinding teeth at the back. If you picture the gaping maw of a habitual carnivore like a great white shark, you will quickly recognise that teeth are shaped according to their function. Like all primates, we have a mix of biting and grinding implements arranged in our mouths. But the relationship between tooth form and food form is not perfect, so palaeoanthropologists have looked for more direct evidence of meat consumption in our ancestors. There is another dimension to the human–animal relationship, beyond the caloric benefit of eating meat, which we can identify in the archaeological record: the nasty diseases we can pick up from animals. These diseases are called 'zoonoses', and they're an integral part of our biological history.

In 2009 researchers from Italy announced that they finally identified the condition that caused the warped spine of the 2.5-million-year-old *Australopithecus africanus* fossil 'Stw 431' from Sterkfontein, South Africa. Two and a half million years ago, this small hominid ancestor may have picked up an infection of a Gram-negative *brucella* species bacteria. This would have caused standard infective symptoms: fever, chills, loss of appetite, headaches, fatigue and muscle pain – but as the bacteria spread into other tissue systems, more devastating changes would have taken place. In a proportion of infections, brucellosis invades the bones and joints of its host, eating away at the critical mechanical junctions of the hips and spine. Inflammation – the reaction to the invading bacteria – causes its own cascade of reactions in the bone. In the lower spine, the nicely stacked, spool-shaped vertebral bones start to develop lesions that erode the supporting bone, leading to squished vertebral bodies shaped more like a soft cheese left out on a hot day and less like solid discs capable of keeping a pain-free distance between adjacent vertebrae. The Camembert imagery is not, actually, a bad start, as the inflammation erodes some of the bone of the spine away and new, reactive bone is formed in response to the infection. If you imagine your spine as a stack of Camembert roundels left out on in the sun for a while and then left to harden, you have a fairly unappetising picture of what brucellosis infection can do.

While unfortunate for the late Stw 431, it's difficult to see how one more infection here or there can be critical to understanding human evolution. But brucellosis is not a human disease. It is, rather specifically, an animal disease. There are several species of the brucellosis bacteria, with each one living preferentially in a different kind of mammal host. But given the chance – contact with bodily fluids, contaminated milk products or meat – the bacteria can jump into a human host. The brucellosis bacteria carried by cows (*Brucella abortus*), sheep and goats (*Brucella melitensis*) and pigs (*Brucella suis*) are particularly virulent in humans, causing fevers and other flu-like symptoms before potentially attacking the joints and skeleton. The US Centers for Disease Control and Prevention (CDC) warns that abattoir workers are particularly at risk, along with veterinarians and hunters, all of whom might come into contact with the bacteria through contaminated blood or meat. And contaminated blood or meat is precisely what we might suspect in the case of the Sterkfontein australopithecine. Two and a half million years ago, this chimp-sized hominid managed to get brucellosis, presumably by consuming infected meat. If the diagnosis holds, it's a great bit of evidence for the longevity of meat consumption in our hominid lineage – and the antiquity of our potential to pick up the diseases of the animals we eat.

Evidence of brucellosis in the archaeological record after this is extremely faint and far between, despite the characteristic changes to the spine that allow archaeologists to identify the infection fairly easily. The constant nature of the battle between our immune systems and the evolution of infectious agents mean that many diseases hit our species quite hard on initial introduction, before both our immune systems and the diseases themselves slowly adapted to a more profitable hosting arrangement that doesn't necessarily kill off the host before the infection has a chance to propagate. We might not expect to see so much of a less virulent infection. However, about 5,000 years ago, brucellosis reappears in a skeleton from the Early Bronze Age at Jericho, possibly the best studied of the world's early cities, in modern Israel.

Today, brucellosis is still a dangerous disease – remember the CDC warning – but the greatest risk is from unpasteurised milk and cheese products. Many more people consume these than directly handle animal carcasses, and this has been true in the past as well. Around 17 per cent of the skeletons of people who perished on the beach in the volcanic destruction of Herculaneum when Mount Vesuvius erupted in AD 79 show evidence of brucellosis infection, and while we know that Roman cuisine included many fresh milk products, the smoking gun in this case is, in fact, a toasted cheese. A cheese recovered from the site had bacterial forms that looked markedly like brucellosis species.

So, on the one hand, we gain beneficial calories from our 'walking larders' – the domesticates that appear as people start to settle in one place and manage the animals in their environment. On the other hand, by coming into such close contact with four-legged disease vectors, we put ourselves in the crosshairs of a host of opportunistic crossover infections. Infection by external animal parasites (ticks, fleas, lice) and internal ones (worms) is far more likely in close proximity to animals. While the majority of zoonoses are *endemic* conditions,* animals have also been responsible for transmitting *epidemic* infectious diseases, which have far more catastrophic outcomes for our species. Epidemics require specific circumstances to arise, however, so we will encounter them in later chapters in this book as humanity builds up the numbers necessary for serious germ warfare. By and large, the introduction of domestic animals to the suite of human settlements does not seem to have negatively impacted overall population numbers, or even produced much evidence of pathology – diseases like brucellosis are actually very rare in the archaeological record. Conversely, the upside of domestication in the form of calorie access is more straightforward. The reduced mobility associated with agriculture is one impetus for domestication; it's very difficult to follow migrating herds *and* tend the garden. It's clear that as human began to settle, animal management became a much more complex, ingrained part of society than the rather freer relationship between, say, our species and dogs. We can see from the evidence at Aşıklı Höyük that early domesticates were kept within a settlement, and that the demography of the animals present was strongly affected by human preference. But, true to form, we *Homo sapiens* couldn't quite let the domestic equilibrium rest.

In many different areas of the world, we have evidence of a new way of life that would open up new areas of the planet to human habitation – and new pathways for disease in us. This has been dubbed the 'Secondary Products Revolution'. While it might sound like something you'd see advertised late at night on TV, this is in fact one of the most game-changing innovations in human history, right up there with agriculture or domestication of animals for meat. The Secondary Products Revolution is a loose term that

* Diseases that occur at a constant low level, and are frequently not fatal to humans.

encompasses the expansion of animal products that people use, mostly from mammals. While human species have hunted animals for quite a long way down their ancestral tree, a hunted animal results in a pretty short-term bonanza. There are the calories gained from eating the animal, the potential of using skin and sinew for clothing or utility items, and of course teeth and horn for jewellery and decoration. However, this is a one-time deal per animal; if you want another meal or another layer, you'll have to go through the time and effort of the hunt again. But what if you just kept the animals close by, and didn't kill them straight off? Aside from cutting down on the 'hunting' part of hunting, you'd have access to animal milk, a constantly re-growing supply of animal hair and four sturdy legs to carry it all on.

Andrew Sherratt, who coined the phrase 'Secondary Products Revolution', maintained that there was an inextricable link between this expanded repertoire of food and other useful products and the social and economic mechanisms needed to manage new, tradeable goods. Additional resources support growing complexity, and to paraphrase his explanation, new foods require new ways of feasting. It takes several years to grow draft animals to full strength, and even animals used for their wool or milk require considerable investment. Similarly, in this model the skill necessary to manage herds used for milk, wool or other products are sufficiently specialised that the pastoralist lifestyle becomes feasible. Specialised herders in areas that are less well suited for cultivation fall into interdependent economic relationships with communities of agriculturalists, and it might be that this sort of division of labour partly spurs the development of the very first 'true' urban centres. The argument has been made that this sort of investment in specialised production is only possible when you have a concentration of people, and potentially a corresponding elite who are able to manipulate the labour and time of others in order to exploit these more demanding goods.

When did the Secondary Products Revolution begin? And more importantly, what did it do to us? The answer to the latter can be traced through archaeology, and gives us an idea as to the former. Actual depictions of milking, plough-pulling and other clear signs of the Secondary Products Revolution do not appear until around 4,500 years ago in Mesopotamia, but did it really take 5,000 years to move from snacking to shearing? As with the process of animal domestication, zooarchaeologists can identify changes in the age structure of domesticated species that might accompany a changing use strategy. A larger number of older animals found at a site might indicate that animals were being retained for the secondary products they could provide,* and an uptick in the number of very young animals might be a clue that people were purposefully encouraging the animals' mothers to continue producing milk. Our evidence for these patterns of animal management is patchy across time and space and has led to considerable debate, so archaeologists did what they normally do and adapted another scientific discipline to address the issue. Organic residue analysis is a means of extracting the (very) trace elements left behind in the archaeological record, for instance at the bottom of ceramic pots. Analysis of lipids – molecules that are found in many fats and other natural sources – can distinguish between the residue of milk fat lipids and the traces left behind by the fat from animal meat. And in the bottom of a variety of 8,000-year-old pots from sites clustered around the Sea of Marmara, we see the traces of well-past-its-sell-by-date milk.

The ability to digest milk into adulthood is a relatively unique human adaptation, and only about 35 per cent of adults in the world today can indulge in milk-based products such as ice cream without a range of unpleasant digestive consequences. Milk tolerance – or more properly, the ability to break down milk by producing the digestive enzyme lactase – is also a very recent evolutionary leap in our species.

* Because they certainly would not be nearly as good to eat.

The genes that allow adults to continue producing lactase are absent in the DNA of samples from the European Neolithic and Mesolithic, and estimates based on DNA place the genetic shift towards 'lactase persistence', or adult milk-drinking, as occurring only about 7,500 years ago. The mutation is far more common in Europeans (and nearly absent in much of Asia), suggesting that it was strongly selected for in some regions and not others. Recent genetic research has shown that a *different* type of lactase persistence arose independently in parts of Africa where livelihoods depend on cattle, which argues that the mutation is very strongly related to positive selection – that milk, in certain situations, does indeed do a body good.

Milk appears to have been very important for certain groups of early farmers. Dairying, and presumably the ability to consume dairy products, seems to have spread into Europe around the same time as the rest of the suite of Neolithic inventions, including animal domestication and farming. Access to milk means access to extra calories, or the potential to have milk available or cheese stored as a fallback food if crops fail. Likewise, the labour-saving potential of using animals for draft (to haul ploughs or drag sledges and even eventually to pull carts and wagons) cannot be underestimated, even if our evidence for the origins of animal labour is currently less obvious than that for dairy. What is abundantly clear is that, collectively, the earliest of permanent human villages adopted animals into their newly settled lives, and, despite the implications of the word, sedentism rapidly spread across large areas of the world. Most animals (pigs being the possible exception) seem to have been domesticated once and then constantly cross-bred with wild types, creating a baffling tangle of genetic traces that give ancient DNA specialists theories to argue about for the foreseeable future. Like very early animal management, there seems to be a net gain for humanity in taking domestication a step further towards using animals for milk, wool and draft power. But again, like every other one of our amazing innovations, there's a price to pay.

At the other end of history, in the mid-nineteenth century, the small Mediterranean nation of Malta was host to thousands of soldiers who had fought in the Crimean War. As a British Crown Colony, this small island suddenly found itself inundated with young northern European men, and hospitals began to report a new disease called 'Mediterranean fever': a series of off/on chills and sweats followed much later by a mysterious sort of rheumatism that seemed disproportionately to affect the servicemen. The British medical establishment of the island suddenly took notice of this new threat to their soldiers, and by the 1880s the infectious agent – the brucellosis bacteria *Brucella melitensis* – had been identified. However, it took 25 years before the vector for human infection was established: about 1 in 10 of the milking goats on the island carried brucellosis, and were merrily transmitting it to the soldiers in the traditionally goat-based milk, cheese and even ice cream. The establishment responded by banning the consumption of fresh milk products and introducing UHT or dehydrated milk, which cut down cases considerably but by no means eradicated them – soldiers kept sneaking off to have real ice cream. Eventually pasteurisation removed the threat, as it has elsewhere in the world; in the US, brucellosis infections plummeted after pasteurisation was introduced after the Second World War.

On Antikythera, I'm relatively safe from brucellosis now. The type of brucellosis carried by goats and sheep is usually passed on through milk products, and there are no milking herds left on the island. Instead, we have the rather fantastically sickly-sweet 'NOUNOU' brand of condensed milk to add to our coffees. Myronas may still be in danger of contracting the disease from goat carcasses, but he's been hunting and preparing goats his entire life and shows no ill effects as of yet. This is perhaps the key lesson from Antikythera. As the population has fluctuated between a few hundred hardy farmers coming together in small villages and today's rather more scattered, pension-drawing last inhabitants, their relationship with their animals has changed. Because there are no longer enough people living in Antikythera to make the time and effort of managing a milking herd worthwhile, the risk of picking up 'Mediterranean fever' has been reduced dramatically. The collapse of such an important part of the traditional Mediterranean diet on Antikythera would bother me rather more if it wasn't for the fact that, despite what my family claims to be a largely European heritage, I do not seem to possess a great ability to digest lactose.* I cannot, however, shake the general feeling that, be it from above, or on the plate, those goats are eventually going to get their own back.

* There's a great variety of ways to be bad at digesting lactose; many people without the lactase persistence gene can still consume milk products without much in the way of ill effects, while others can't do milk or ice cream but might be able to get away with cheese or yogurt. I personally still consume all of these things, but, again, ill-advisedly.

CHAPTER FOUR

Revolution

The year is 2008. It's a Thursday, around mid-afternoon, and the smell of charcoal is heavy in the air. Levant, Thursday barbecue master and sometime site guard, has assumed his traditional position just outside the kitchen, half-barrel grill dragged into pride of place in the courtyard gravel and just beginning to smoke. Hand-formed minced meat skewers have been laid out on an industrial scale on the table beside him, the big flat metal blades of proper Turkish kebab skewers set into meatball hilts still bearing the undulating pattern of the fingers that pressed them into shape. A particularly persistent fly is re-enacting the opening scene of *Once Upon a Time in the West*. Archaeologists are milling about like stray cats in a fish restaurant, slinking around the porch posts and railings and positioning themselves with wide, pleading eyes as close as possible to the kebabs. There are nearly 120 archaeologists on site, housed in bunk beds in the dorms lining the quadrangle of the excavation house, or littered across the sun-baked fields out back in a riot of mismatched tents, ranging from the palatial striped pavilion (ceramics team) to the poorer, disintegrating, accidentally skylit dome (that would be mine).* Earlier in the day, the pickup from the Efes brewery came in on a slow-moving brown haze of dust and hope, kicking up a cloud in the parking lot and disgorging the standard beer delivery plus

* Some enterprising folk maintained two residences on site; one now-established scholar for instance kept a tent expressly for (ahem) facilitating sociability. I always preferred tents for the exact opposite reason, but the downside to having your own canvas bubble was the fact that they get to about 45 degrees by 7.00 a.m., and you might occasionally wake up to find your guy-wires crossed with a tent erected for purely (cough) social purposes.

whatever branded material the site consumption level has merited by this point – previous highlights include a free refrigerator and an entire suite of Efes patio furniture. One of the students from Cambridge has demanded (and more surprisingly, got) my second-best white button-down shirt, no one has seen the conservation team in hours and the supervisors have all mysteriously vanished with half the glassware and all the ice. We're on the cusp of a ritual that has been played out on this otherwise relatively unremarkable scrap of the Anatolian plateau for decades: the Çatalhöyük weekend. It's an occasion marked, like every other party in pretty much the entirety of human history, by feasting, by group consumption of the most highly valued goods available (kebabs and beer, in our case) and by ritual (more on which later).

Turkey is a good place to stop and take stock of the Neolithic Revolution. It's not that the adoption of sedentary, agricultural lifestyles is unique to the region – there are one or more Neolithic horizons on every continent bar Australia and Antarctica. The countries to the south and east of the Turkish heartland of Anatolia have much earlier evidence for the adoption of Neolithic lifestyles. But what is very interesting about studying the Neolithic in Turkey is that the vast plains of Anatolia – ludicrously hot in summer, bitingly cold in winter – are one stop up the road from the Levant and the Fertile Crescent, where these revolutions were kick-started. Anatolia is not the place where the Neolithic revolution started, but it is the point of first contact for what became an epidemic of cultural and demographic shifts. One of the most vexed questions in archaeology is the one that Gordon Childe thought he'd answered nearly a hundred years ago: what drove this Neolithic Revolution across thousands of miles and thousands of years, obliterating the hundreds of thousands of years of life that had gone before?

The success of the Neolithic is the success of its transmission. It begins as a regional phenomenon that catches in all corners of the earth, springing up *sui generis* multiple times, blazing forth with slightly modified seed heads and marginally sturdier housing, only to sputter and collapse in the face of as-yet poorly understood pressures: ecological, demographic or perhaps even social. Often, the conceptualisation of human history lazily carries this analogy forwards, taking the case of the Near Eastern Neolithic 'suite' of adaptations and accusing it of spreading like wildfire into Europe and beyond: a blazing path towards progress, civilization and the sort of society that publishes books by universalising philosophers who can't quite picture themselves anywhere else but the pinnacle of human achievement. But at some point we have to step back from this inflammatory arc of progressive thought and examine how the world came to embrace sedentary, agricultural life.

The Neolithic changed people. Beyond the impressive cultural and organisational rearrangements, beyond the lifestyle and social aspects of settled, farming life in different places and times across the globe, the Neolithic Revolution made actual physical changes to its adherents. The scars of the revolution are writ in our smaller frames, weedier legs and shrinking faces. Our teeth have holes in them from the

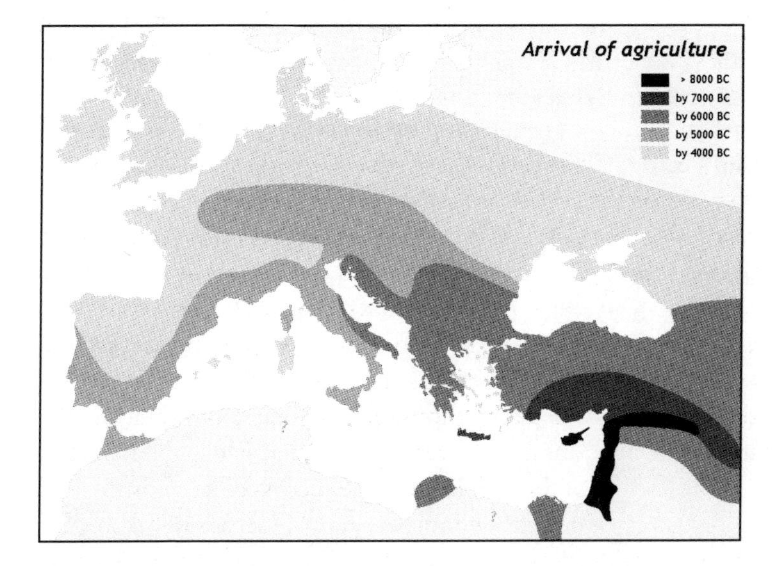

Figure 3 *The spread of agriculture. Source: Andy Bevan.*

caries-causing agricultural products we gleefully consume, and worse still, they don't even fit in our mouths very well anymore. The teeth of early adopter agriculturalists carry the fossilised record of their repeated illnesses and growth disruptions. Bringing animals into our lives and homes brought debilitating, spine-crunching diseases, and yet the fad for sedentary, agricultural lives spread rapidly across the globe. If you want to benchmark that speed, consider innovations like stone tools that cut on both edges: the Acheulian industry that followed *Homo erectus* all the way to East Asia took nearly half a million years to make inroads into Europe,* against the handful of millennia it has taken for agriculture to spread to all corners of the globe.† The question we have to ask is: why? Why did this revolution occur, and why was it such a runaway success? If the Neolithic Revolution was so terrible for our bodies and our survival, why has nothing else we've ever done as humans proven as popular?

Theories of a species-wide tendency towards self-extinction aside,‡ there are a few explanations for why the Neolithic caught on. One in particular explains not only how it caught on, but what might have jump-started it in the first place. There have been two main schools of thought on this. The 'cultural diffusionists' argued that it was the *idea* of the Neolithic that spread across the globe; that the technology of farming was sufficiently desirable to be picked up by all and sundry as they came across it. This was a riposte to supporters of a 'demic diffusion' model, who equated pots with people and saw signs of Neolithic lifestyles in the archaeological record as a clear indicator that it was the farmers themselves who had expanded their territory. With the advent of next-generation sequencing in ancient DNA

* To be fair to our smaller-brained ancestors, it was very cold.

† And beyond; the first seeds in space were sent on a 1942 V-2 rocket trip, and current crops aboard the International Space Station include romaine lettuce.

‡ Though not too far aside. Maybe 4° C (8° F) or a few H-bombs aside.

research, bioarchaeologists now have new tools to bring to bear on this debate; we can start to look at where all of these Neolithic people came from, and where they went.

Rapid advances in aDNA technology mean that we can take vast amounts of genetic data and compute variations and divergences between individuals to build up a branching tree of likely lineages. Ancient DNA research has moved from the shorter sequence of information encoded in mitochondrial DNA to more complicated chromosomal DNA, with ever-increasing sophistication of computational and capture techniques. This allows researchers to access even more detail on the inheritance of very subtle shifts in DNA marking out different 'haplogroups' – groups that have inherited the same sets of tiny alterations to their genetic code. DNA testing of living populations and archaeological remains has very much thrown the (undomesticated) fox among the (domesticated) chickens: folded tightly into the tiniest scraps of carefully retrieved DNA are signs of large population movements at key moments in our species' history. In Europe, where these questions have been most thoroughly researched, we see evidence that haplogroups from early famers are markedly different from those belonging to the longer-established hunter-gatherers of the continent. Parts of Europe still have a high quotient of this pre-Neolithic material – the Basque Country, Finland – but research has identified commonalities between the early farmers of Anatolia and subsequent famers of Europe. While it is Sardinia that is home to the closest genetic match to those early Anatolian farmers, even inhabitants of Neolithic Ireland, out on an island on the entirely opposite side of the continent, show clear markers of being part of the same Near Eastern ancestry.

The evidence from ancient DNA is of course hugely complex, and without genetic material from long-gone lineages we may never be able to plug all the gaps in the Neolithic story. Indeed, in the last few years aDNA research has thrown spanners into the works of several theories of human population genetics. We have learned that out-of-Africa modern humans interbred with not one but several species of extinct humans, including Neanderthals and the mysterious Denisovans, and we have discovered that nearly half of the material that makes up much of modern Europeans came from the Eurasian Steppe as recently as 5,000 years ago – so not from our Near Eastern farmers. The story of the Americas, and of Asia and Africa, is still waiting to be told in as much detail; but there is good money on it further upending our delicately balanced scientific apple cart. In India there is now evidence to suggest the movement of haplogroups otherwise found in the Near East and Eastern Europe through the northwest corner of the subcontinent, following the path of early evidence for agriculture. These groups may be an entirely different group of people from the farmers who spread into Europe, expanding out of the Near East in wavelets alongside people who may well not have been full-time farmers.

A very recent, late-breaking news story appeared in summer of 2016 to further complicate our ideas of how farming – and farming people – spread across the world. The ancient genome of people who lived in the Zagros Mountains at the very dawn of agriculture was sequenced by an international team led by researchers at Johannes Gutenberg University in Germany. The genomes revealed that bodies found in a cave site dating back to around 9,500 years ago (the time when agriculture really took off in the upland native habitat of the wild species that would go on to become our domesticated plants) were not from the same population as the farmers who spread their DNA from Anatolia up into Europe. The Zagros DNA showed no evidence of mixing or contributing heavily to the wave of agriculturalists who swept up out of the Near East; rather, their DNA signature spread to modern-day Pakistan, Afghanistan and particularly into the Zoroastrian population of Iran. The mountain folk of this study lived just one mountain range away from the Anatolian farmers of Çatalhöyük and Aşıklı Höyük. By the Neolithic period the regions in which they lived were drawn into an ever-increasingly interconnected regional network that drew in raw materials like obsidian from the same places. Their genes, however, testify to a split with western Anatolian farmers some 30,000 to 40,000 years before.

This kind of evidence of genetic transmission argues that there might just be something worthwhile in 'seeing people in pots'. However, further evidence that the farming transition was not just a simple case of farmers and their DNA spilling out of the Near East is piling up, as new studies hot off the presses reveal that it wasn't just mountain men and plains people occupying the Fertile Crescent at the dawn of the Neolithic, but other, separate groups such as the Natufian farmers from further down south in the Levant, who would intermingle with Anatolians but later go on to leave their DNA in East Africa.

Likewise, the other major advance of the Neolithic – the domestication of animals – seems to be spread across very different genetic groups. The people of the Zagros Mountains from 10,000 years ago may have lived next door to the world's first farmers, but they show closer genetic affinities with earlier European hunter-gatherers than with the Anatolian agriculturalists. Their way of life may have depended more on goats than on plants, so perhaps their 'Neolithic' was a rather different experience than the one that spread up into Europe. We have assumed for a very long time in archaeology that there is only one right way to 'have' a Neolithic, and we have based it on our best-known scenario in the Fertile Crescent of the Near East. The devil, however, is in the details. There may be many ways to have a Neolithic. It might be that we are simply unable to catch the archaeological signs of some of these different approaches because the plants and animals involved are the same, so we assume that they were exploited in the same way. While we see some evidence from the trace elements left in bones and teeth that some non-farming groups had a diet with a similar proportion of plants as agriculturalists, it might be that we simply don't really know enough about what hunter-gatherers or pastoralists or early agriculturalists ate to tell the difference. It may be better to imagine the process of cooking up the Neolithic (at least in Europe) as using a series of diverse ingredients added in different amounts and at different times. Simultaneously, you'll need to imagine that someone keeps pulling the thing from the oven and taking out specific amounts of flour or egg after it's started cooking. The whole model[*] turns out to be … half-baked.

These huge introgressions of people (and their pots) have rather large repercussions for how we model the spread of new adaptations like farming. Childe's concept of diffusion saw the Neolithic spread as an idea whose time had come – perhaps owing something in this to his interest in Marxist thought and social progress. But what we see in the archaeological record is far more complex than a linear march of one particularly successful idea or genetic group. The constant movement of people and genetic expansions, mixing and admixing, the to-and-fro of waves of humans between Africa, Europe and Asia through the tide pools of the Near East, leaves a strange residue of Neolithic populations and cultures behind. It's just as unlikely that these waves of humans and cultures existed in total isolation and had no effect on each other as the antiquated idea that there is only one way to discover America.[†] Our problem as archaeologists isn't so much that we don't follow the principle of Occam's razor; it's that we can't even begin to see the shape and size of the possibilities it's cutting down.[‡]

One of the most frustrating aspects of archaeology is that it's slow-burn science – eureka moments and world-shaking revolutions in human adaptations are not punctuated events. Unlike the other sciences, where your best results can literally go bang in an instant, what archaeology finds is a slow succession of incremental changes. Or, at least, we find bits of those slow changes; the rest are down gullies, at the bottom of lakebeds, under the sea or up a mountain. Human evolution is *boring*, and the changes we have been talking about – settling down, farming, domesticating animals and slowly spreading over the earth – took a long time. It's a mistake to see the

---

[*] Much like my skills in analogy.

[†] For instance: early, using the Bering Strait.

[‡] This is surprisingly true in physical anthropology. No one expects the Denisovans.

biggest revolution in human history, the Neolithic, as a split-second moment or as a monolith.

Different parts of the world had different Neolithics at different times. Much of what you hear endlessly repeated by lifestyle coaches and nutritional supplement ads is based on our Euro-centric view of how farming and settled lifestyles kicked off. Bioarchaeologists use the markers of stress discussed in Chapters 1 and 2 to judge the impact of these transitions, and we see very different patterns depending on where and when we look. The transition to agriculture based on wet rice farming seems to have gone swimmingly, with little signs of stress in the bones and teeth in Southeast Asia,* while the maize-based diets of the American Southwest seem to have ended in tears.† We say that sedentism led to farming, and that this led to towns and cities, but we know there are exceptions: the archaeological and historical record is full of groups who tried farming and/or settled life and then thought better of it, from the Natufians 14,000 odd years ago to the agricultural refuseniks of the Southeast Asian highlands less than 200 hundred years ago.

Coastal resources in particular seem to have inspired the construction of permanent settlements and complex social hierarchies; we'll see this again with discussions of the Chumash and other Californian indigenous groups in the next few chapters. Having just watched a former compatriot from University College London, the British Museum's Americas curator Jago Cooper, swan around the Northwest Pacific Coast of North America for his most recent BBC special,‡ I was impressed not only by the usual grandeur of such productions but also by his insistence that the highly complex world of the Native Americans of those coasts was of

---

* Thailand, Japan.

† And cannibalism: see Chapter 8.

‡ In which someone paid for him to go canoeing, kayaking, mountain biking, lumbering, boating, salmon fishing, carving, weaving and god knows what else all across the coastline of Western Canada; that's what is known to UK archaeologists as 'jammy'.

extraordinary duration. While arguments abound about whether the people of the region really were settled year-round by about 7,000 years ago, there is clear evidence for thousands of years of sedentism without reliance on farming.

The cultural continuity of those thousand years of complex hunting and gathering gives us perhaps a better insight into the Neolithic Revolution than all the ethnography that exists of modern hunter-gatherer groups who have been pushed to the absolute edges of the habitable planet by encroaching agriculturalists. Without the interruption of climate, of colonialism, of changing circumstances in the form of population, contact or environment, human cultures can be remarkably parsimonious with their adaptations. Unless hurried on by circumstances beyond their control, human 'progress' works like the mills of God, and grinds exceeding slow.

We can see the torpor of the Neolithic in the sheer length of time our 'revolutions' take. A full millennium, give or take, seems to be about average for native adoption; developing the whole scenario from scratch takes more time, and picking it up from the neighbours takes less. Archaeologist Andrew Chamberlain points out that there is a 1,000-year lag between the appearance of the first domesticates in several North American cultures and the adoption of the intensive farming of the classic triumvirate of maize–beans–squash, and this is borne out in a demographic lag. The increased fertility that is the hallmark of Bocquet-Appel's theorised Neolithic demographic transition does not occur until *after* the shift from what has been called 'low-level' food production by Bruce Smith. Low-level food production occupies the long, drawn-out grey area in many parts of the world where farmed crops play an important (but not exclusive) role in provisioning a society. But, as hinted at in earlier chapters, the key to Neolithic success was … more people.

Archaeology may not always have the immediacy of physics* or some other sciences, but it does slowly build up the evidence

* Or at least, in theory, it shouldn't. Though anyone who's ever been in my car off-roading to site might disagree.

we need for our own eureka moments. The fertility-rise harbinger of our species' enormous swing vote for farming is something we can access, albeit with a great deal of trouble. The evidence of increasing fertility I saw in my work in Central Anatolia, locked into the teeth of children 10,000 years gone, argues for a similarly slow boil at work in the ancient Near East as in the Americas. Occupation at Aşıklı Höyük lasted at least a millennium, with nearby Çatalhöyük occupied for the millennium that followed. That is an astonishing swath of time to get around to sorting out being a village, and it's entirely possible that the foot-dragging nature of our march to urban life reflects a real problem: the Neolithic was dangerous. Not in a 'let's go hunt aurochs' kind of way (though this did continue until we ran out of aurochs), but in a more insidious, long-term-debilitating kind of way.

My recent collaborative research with Emmy Bocaege of the University of Bordeaux has emphasised the dangers of this transition. Using incredibly fine-grained analysis of tooth growth (the 'tree rings' mentioned in Chapter 2), she has uncovered a pattern of growth disruption in the teeth of the children of Çatalhöyük showing that they were consistently exposed to episodes of either malnutrition, disease or some other metabolic stress. Compared to the earlier, less firmly Neolithised inhabitants of Aşıklı, the children of Çatalhöyük appear to have had a harder time of it. They suffered a higher number of incidents that left scars on the dental record of their growth. What must be remembered, however, is that to form a scar, you have to live through the wound: the marks of childhood diseases that we can see in teeth are badges of survival. Just like at Dickson Mounds, a continent and millennia away, we can see in the rising numbers of hypoplastic defects a story of risks survived.

Life at Çatalhöyük may never have been easy,* but it continued for 1,000 years. It continues to this day as part of one of the most famous archaeological projects in the

* In the immortal words of excavator Anies Hassan, broadcast over the site radio: It's soooo hot.

world, a UNESCO World Heritage site that attracts both visitors and researchers to the high flat Anatolian plain year after year. While excavations are now winding down after more than 20 years, there is still a lot that remains to be understood about the Neolithic transition in Anatolia: how people lived day to day, how they ate and what rhythms their societies moved to. Project director Ian Hodder is a firm proponent of the post-processual school of archaeological thought, which tries to take into account the fact* that the things we dig up as archaeology were made by people too – people with individual motivations, social roles and a more complex reality than simply inventing farming by Wednesday and chilling for a thousand years.

Sometime after nightfall on that day in 2008, the scattered community of Çatalhöyük came back together, drawn by the roaring bonfire and the gravitational pull of alcohol and music. First on scene were those who had dedicated their afternoon to meticulously curating playlists of obscure 80s prog-goth and 90s hip-hop classics, to bickering over the exact proper layout of sticks of varying sizes for the purpose of effective burning, and in general to preparing the ritual space. Slowly but surely, the celebrants themselves entered. Çatal has a longstanding tradition of 'themed' party nights, to the extent that it's no longer conceivable that a group of archaeologists in the middle of absolutely nowhere could ever just have a few quiet beers at the end of the working week. Instead, the brief amount of respite given to the diggers has been entirely taken over by the construction of the extraordinary sartorial displays demanded by the light-hearted but increasingly intense costume competition over the years. Slowly, wearing their costumes scavenged entirely from the materials available on site (trash bags, field clothes, duct tape, egg cartons†), some of the brightest and best

---

* Problem.

† Without which I would never have been able to achieve the desired hair height for the French Revolution theme night.

archaeologists in the world trickled out onto the dance floor.[*] On this particular Thursday night, the Cambridge team took the group costume idea even further, waiting to make their entrance until the party was more or less in full swing; suddenly, the hair-metal dance floor filler clicked off on the tinny speakers. Silence descended, only to be rather violently broken by traditional English folk music at top volume, accompanied by a squad of five completely white-clad, stick-waving Morris dancers, high-kicking their way across the compound.[†]

While not the first time that I've experienced elaborate stick-dancing displays in the field,[‡] the Morris dancing particularly struck me: here we were, digging down into the very beginnings of settled human life, but no matter what, people found a way to come together to eat, drink and celebrate. The charcoal from our fire would eventually come to resemble that found on site, not 20 metres (65 feet) away, from fires that burned 9,000 years ago. In the absolute dog days of summer in 2016, I would hear[§] that the Çatalhöyük zooarchaeologist Nerissa Russell presented evidence of disarticulated crane bones found on site – these could have been used as ritual costumes. They could, even, have been *better* costumes than my very last-minute Rambo. In all this time, everything and nothing had changed – from feasting to farming to the mud-brick construction of the houses. At Çatal every small aspect of the Neolithic world we uncovered brought new insight into the ways people had thought and

---

[*] Highlights include: nuns, vicars, gods, goddesses, the conservation team as Ninja Turtles, and a fabulous seventeenth-century full-skirted gown constructed entirely from a patio umbrella and Efes bottle caps.

[†] Which did finally explain what had happened to my second-best digging shirt.

[‡] Traditional dancing is a much-underestimated hazard of archaeological fieldwork.

[§] On Twitter, which is why the tweeting of conferences is wonderful. Thanks Jens!

conceived of their world. The enormous *bucrania*, bits of wild bull skulls with horn cores attached found mounted on walls and benches, were surely remnants of the ritual hunts and feasts of those long-gone early adopters, feasts that left piles of bones and debris in the same way that the Çatalhöyük dance floor would the next day be covered in Efes bottle caps. At Çatal, we can only imagine what the ritual context of these feasts would have been like, piecing together events from the evidence of animals eaten, hearths lit, floors cleared and buildings or special areas set aside.*

At sites like Göbekli Tepe or Musular, hunter-gatherer sites in the same region dating even further back in time, we see the power of place in drawing people together. These were focal points in a much wider landscape of human occupation – bringing together dispersed groups for a brief period of communion. At sites like Aşıklı and Çatal, we see the same tradition of feasting and communal life, but it appears that no one was willing to let the party break up. Here the solid walls and established hearths reinforced the idea of a community settled and separate from any other groups still wandering the landscape. This new kind of identity – the town, or village – is the beginning of a trend that would sweep the rest of the world along before it on a rising tide of tension and conflict. The people of Çatal found a new way of living, one that would see population boom and eventually force new ways of living together to be invented just to keep all those people engaged in this Neolithic experiment. The pressures of a rising population ripple out into the environment as more land is put under cultivation, more animals are hunted past the point of sustainability and more people share the same space.

Luckily, people are very good at adapting to the messes they make. The revolutionary changes of the Neolithic, however, could not be completely patched up by a bit of community solidarity, bought with some fireside singalongs and wild bull meat, any more than drunken costume parties

* On balance, Morris dancing probably did not feature.

could truly patch over the personal, professional and pop-cultural disagreements of 120-plus archaeologists at Çatal.* The myriad needs of a world increasingly full of people, forced into sharing their space and time, would push the early towns and villages of the Neolithic into the social complexities of the first true cities. Ever more complicated ways of balancing the needs of individuals and the needs of the community as a whole would become an actual job, and once you've invented managers, it's no simple thing to return to the easy egalitarian world of our hunting and gathering past. Of course, our idea of egalitarianism in the past might be just as outdated as the Çatal playlist – something the next chapter will have to look at in detail. But in a world where farmers were everywhere, the only thing left was to try to survive the change, and, as our world became increasingly more urban, that became harder and harder to do.

* I am still peeved about the crossed guy-wires thing. Ten years later.

CHAPTER FIVE

Power of Equality

There are two revolutions to get through in discussing the effect urban living has had on the physical health of our species. The first, the Neolithic Revolution, has been described in the previous chapters. The Neolithic of the Near East spread out into North Africa, and then through the Balkans and Central Europe, eventually pushing up into the very farthest regions of Northern Europe, over a period of about 3,000 years. The Neolithic of the Americas seems to have stopped and started independently in a number of places over a few millennia – from the domestication of potatoes and llamas in the south to grasses and legumes in the north. The African Neolithic is in fact plural, and occurs in multiple places and times with various types of millet and sorghum domesticated alongside imported Asian crops, and of course the critical Ethiopian domestication of coffee. In Asia, a wide variety of Neolithics (based on rice, wheat and tubers) similarly rose (and failed to rise). The story continues with mini-domestications and more successful long-term efforts occurring across the globe, from the very early domestication of the banana in Papua New Guinea* to the very late taming of the vanilla orchid in the last 500 years. While many of these eventually took hold and cemented people in place, the process of studding the global landscape with villages and sedentary populations took a considerable amount of time. However, in certain areas of the world, it seems that these fixed villages full of people rapidly began to build up numbers, accreting more and more

* Which, in a style typical to the region, fails utterly to conform to expected notions of agricultural development; there is evidence for domestication of banana, taro root and the yam around 6500 BC, but the practitioners successfully avoided a sedentary village lifestyle for more than 2,000 years after this.

inhabitants at strategic locations until, for the first time, we begin to be able to talk about 'cities'. This is Childe's second revolution, five millennia down the line: the Urban Revolution.

It's difficult to pick a single definition for the concept of 'city', and even more difficult to choose how to identify 'urbanism' in the archaeological record. 'City', in English, is a word that comes down from Latin via French, but in a fairly telling error, the *civis* of origin is in fact the inhabitant; it's *urbs* that originally denoted the place. So perhaps we are already primed to recognise a city as such based on the nature (and number) of its inhabitants. As a child, I was told that a city has a cathedral; this can be fairly easily dismissed as a rather parochial interpretation when China builds urban conglomerations for 15 million people without so much as a bishopric. The distinction between city and not-city may seem obvious in the modern world, but it's in fact very difficult to try to extend it back into the distant past. When do we start living in cities? What is it about cities that make them different, that distinguish them as an environment?

Gordon Childe could never shake free the idea of cities from the idea of civilisation. For the progressive Childe, who firmly believed in the linear nature of human achievement, the beginning of civilisation was the city, and the *civis* and the city were more than just linguistically intertwined. His checklist for recognising a city included size, specialists and the concentration of wealth, all recognisable in the archaeological record through great pieces of monumental architecture. But these things show up in other contexts as well: the monuments of the hunter-gatherer ritual site of Göbekli Tepe, mentioned in the previous chapter, or the specialist 'big men', shamans and tool makers of many non-settled societies. The characteristics that seem to hold are actually a combination of size and settled-ness, as well as evidence of social stratification that lasts generation after generation: a density of population wrapped up in some sedentary, aggregative activity that leaches goods and people into the service of a larger (frequently, temple-shaped) monumental structure.* Monuments, and large public spaces where a community might gather, we have seen before. In the Ukraine, the largest Neolithic sites of the Cucuteni-Trypillian culture may have held tens of thousands of people, but scholars have argued they show not a trace of evidence of differences in rank. As archaeologist David Wengrow has suggested, it might be that these Ukrainian mega-sites don't have to mean year-round, settled lives – for instance, there doesn't seem to be the type of big impact on the local plant environment you would expect had the sites been surrounded by cultivated fields. Certainly, there are ethnographic (and contemporary) accounts of migratory, agro-pastoralist or simply pastoralist cultures in places like North America and Mongolia that bring together huge numbers of people in heavily stratified political and social circumstances for seasonal gatherings, without a single permanent shelter to mark the occasion.

The special classes of people that we can clearly identify in the archaeological record before the advent of organised hierarchies are often craft specialists or what are commonly termed shamans, individuals who went into the earth (and were later retrieved from it) accompanied by key signifiers of their personal status.† The burial of an infirm old woman with some hundreds of tortoise shells in Hilazon Tachtit cave in Israel in the Late Natufian period, early in our experiment with sedentism, marks the death of an individual of some ritual or social significance, but she is alone with her tortoises. There are no other tortoise-full graves, no long tradition of a Tortoise Queen; and, most significantly, there are no young Tortoise Princesses. This is what, for anthropologists, signifies the boundary between 'egalitarian' societies, where

* To needlessly paraphrase Douglas Adams in defining monumental architecture (he meant an airborne party, but we'll elide this detail): it's pretty recognisable as one hell of a thing to be hit with in the small of the back.

† This status is assumed to be ritual if we cannot figure it out: First Law of Archaeology.

all categories of people are more or less the same, and stratified ones. In an 'egalitarian' society, the status differences between individuals, or even categories of individuals such as male and female, are in-life differences that are *ascribed*. They are earned or given in life, not before birth. Structural social inequality is recognised (with the sort of caveats and exceptions beloved by all social scientists) by the inheritance of social position. At Hilazon Tachtit they may have a Tortoise Queen who has earned her rank, but there are no Tortoise Princesses born to theirs; this is a society for whom it is not, in fact, tortoises all the way down.*

Egalitarian, of course, never actually means what you think it means. While in concept, Bob the hunter-gatherer has the same inherent social standing as Frank the hunter-gatherer, everybody agrees that Frank tells much better stories and that Bob is a bit of a liability when the psychoactive religious rituals come around. In an 'egalitarian' society, the esteem (and, perhaps, extra bits of honey or meat, special beads, fox teeth or even tortoise shells) Frank is accorded later in life is essentially awarded meritocratically. None of this matters, of course, to Barbara, who would be fantastically good at telling the story about the Creation Eagle or what have you, but there are no speaking roles for women at the

* One of the best apocryphal stories in the history of science is that of a little old lady, listening politely to a Great Professor give a grand public lecture on the nature of the physical universe. After finishing his talk with a rhetorical flourish to silence the most obstinate of critics, the Great Professor proudly calls for questions from the audience. The audience, stunned by his grandiloquence and the force of his arguments, is silent, bar the little old lady. She stands up from her seat in the back and congratulates the speaker on his delivery. 'But,' she adds, 'of course you know that you're wrong. Everyone knows the world stands on the back of a giant turtle.' The Professor condescends to this unexpected heckler with a knowing wink to his audience of learned men, and asks the little old lady, what, then, does the turtle stand on? The little old lady shakes her head at the folly of great and learned professors, and sets him straight: 'It's turtles all the way down.'

Magic Mushroom Ceremony. Arriving at the understanding that 'egalitarian' as observed in living hunter-gatherer communities was probably as much a misnomer in the past as it is today took anthropologists a long time, and seems to coincide with the world creakingly slowly learning the word 'intersectionality'.*

Status, of course, matters very much. While sometimes it may only mean a difference between table assignments in Los Angeles restaurants,† or a spot on the leader board for tortoise-burial, it has had pernicious and even deadly consequences for much of human history. While the blandest definition of status refers to any position in a social network (age, family relationship, etc.), the sort of status that has the greatest effect on individual humans in the past is the social status that determines access to resources. Power *to*, if you like Hobbes, or power *over*, if you prefer Weber. We see the first archaeological signs of differences in people's ability to acquire food, labour and goods arising in the Neolithic.

Prominent UK archaeologist and former director of the Institute of Archaeology at University College London Stephen Shennan points the finger squarely at the advent of farming as a precursor to the idea of 'owning' land and the rise of emphasis on material wealth as a key factor in reproductive success – a process so embedded in our modern world that it formed the basis of economist Thomas Piketty's now-famous classic economic history just a few years ago.

---

* This also perhaps explains why popular children's books about the dawn of humankind tend not to feature women until at least the Neolithic, when they can be safely imagined quietly grinding something. According to archaeologist Matthew Pope, in the notable exceptions where early hominid females are pictured, they are exclusively allowed to wear 'pensive' as an expression. Presumably this is the ancestral version of 'selfie duckface'.

† Perhaps the most strictly delineated hierarchy known to humankind. Your host *knows* the box office on your last film – if you don't want to be sat by the toilets, tipping is advisable.

Using Eric Smith and colleagues' three categories of wealth that can be transferred from parents to children – investment in physical form (embodied wealth; think of this as making sure your kids eat right), in social networks (relational wealth) and finally in actual material goods that can be handed down the generations – Shennan identifies the Neolithic as the point at which that key material wealth becomes a place, a defensible territory. He likens this to the creation of an ecological niche.

Other researchers have suggested that the accumulation of material resources might have taken other forms – that 'wealth' could have emerged through the manufacture of high-value objects, or by building up animal herds that could be handed down the generations. The famous demagogue of data-driven archaeology, Lewis Binford, would take issue with the idea of villages as the only place social complexity ever happens, as indeed would most anthropologists if archaeologists ever asked them. In case after case among extant (or recently extant) hunter-gatherers, he argues for considerable status hierarchies; take the example of the complex cultures of the Pacific Northwest of America that Jago was mountain-biking and canoeing around, or the Chumash culture of California, which will be our constant companions in this book thanks to the geography of my early life. Both had highly developed internal hierarchies where status could be inherited – the Chumash even had pierced-shell industries to mint shells that essentially functioned as money – and both relied on the sea. Depending on aquatic resources seems highly correlated with these kinds of complex social arrangements. One argument that reconciles both Shennan and Binford's stipulations about wealth and ownership of land is that certain types of seafood make for a particularly immobile resource. Abalone cling to rocks, so the communities that eat abalone cling to the land outside those rocks just as tightly, utilising whatever political systems they can to keep within range.

Recent research suggests that just as ownership of land became increasingly complex, management of animals also became specialised, with different strategies for maximising milk, meat or surplus animals – methods of investment designed to minimise risk that could have led to embodiments of wealth and further social complexity without having to set foot in a city. Anthropology has largely debunked that linear, progressive narrative of increasing 'complexity' reaching an apex in agrarian states; we know that humans will find a way to complicate their social structures no matter how they make a living. However, there is still something very special about those fixed points where populations aggregate. Farming requires a captive audience, or perhaps more accurately, making a living from agriculture pins people to the landscape.

The anthropologist James Scott has argued that fixed-field farming – where people are dependent on one particular area of land for cultivating enough food to feed themselves – is, in the immortal words of Admiral Ackbar, a trap. It creates a literally captive population, tied to the land and therefore stuck in whatever power structure happens to reign. If we accept that agriculture does tend to up the birth rate, what you really sign up to with the farming package is a cycle of intensification: more people means higher population density, which in turn means that you cannot easily just up and move your farm – there will already be somebody there. As Scott put it, it's the grain that domesticated us, and not the other way around; farming bred more humans to grow more grain, with very little concern for our wellbeing.

One theory about human social complexity, constructed by anthropologist Robin Dunbar but frequently referenced in a world of easy digital 'friends' and 'followers', suggests that at some point the number of people around really outweighs the ability of individuals to organise their one-on-one relationships with the rest of society. This might be an opportunity for more complex social hierarchies to sneak in. 'Dunbar's Number' is the number of people he theorises we are able to retain enough information on (and interest in) to maintain a close relationship, based on a series of studies of primate group size and social organisation; the number itself is calculated to be around 125 in humans. While I'm sure that the author himself would acknowledge this to be a plastic concept, there is something slightly irksome about the assumption that human socialisation is a monolithic, one-time evolutionary by-product, and that our extensive human experiments in living in bigger or smaller groups wouldn't alter the way we learn to be social. If Boris Johnston, former Mayor of London and current UK Foreign Secretary,* can not only identify (in gym kit and from across the street) a higher education policy maker he met exactly once some years previous, as well as remember her name, one can safely assume that some people are able to hold on to the details of far more than 125 people.

What all of these models have in common is the clear signal that it's *people* who are the problem – tied to the land, accumulating advantages through new forms of wealth and building up numbers. Archaeologists and anthropologists have argued that the advent of farming and settled societies was a key turning point in establishing inequality in our species, but it's difficult to decide which aspects of the archaeological evidence mark a clear trajectory towards social differentiation. The Natufian Tortoise Queen of Hlatchit Tizalit cave was clearly differentiated from other nearby burials in terms of grave goods, but is that enough? Other burials of the period occasionally have small, special objects included; are they personal mementos or should we identify the Natufian as the start of the inequality that would come to play such a huge part in determining whether we live or die? The much-later Central Anatolian sites of the Pre-Pottery Neolithic, Aşıklı and Çatalhöyük, which have been discussed previously, have their own evidence to give. At Aşıklı in particular, there are no signs of the personal goods or gods that might mark out individuals or households with different access to resources, be they material or meta-physical. Large communal storage and communal spaces with traces of a central hearth, ample seating and remains of

---

* And living proof that a Classics degree in the wrong hands is a dangerous thing.

collective activities like feasting testify to a much more egalitarian mode of existence.

Slowly, however, the communal spaces of the earliest villages get roofed over, and the passage between the personal world and the communal world gets narrower. We see this physical shift in the very nature of buildings themselves, as the great collective communities of the early agricultural experimentalists start to reshape themselves to meet the demands of a growing population. Individual hearths inside individual buildings replace the old ones that sat outside, open to all. Communal storage gives way to individual hoards, the surplus potential of agricultural wealth suddenly something to accumulate and protect. The big public spaces for feasting and rituals that tie the community together shrink, or are slowly elevated out of reach of the commons. We see doorways on temples and hidden rooms where the gods are available only for private conversations with a select few. The old communities of the early Neolithic disband, fall apart or move away. New sites are founded, this time with walls. The comedian Eddie Izzard nicely summarised the sum total of archaeological discovery as 'a series of small walls'; as unimpressive as they sound, they are our only tangible evidence of the changing spaces in which humans lived.

What price inequality? Biologically speaking, does it really matter if a little hierarchy creeps into the world? In fact, it would be hard to identify a factor that has had more impact on the trajectory of our species. Inequality today is the defining factor in how long we live, how healthy we are and even how we grow. A little boy growing up in the most affluent communities in modern America, for instance, can expect to live 15 years longer than a boy born into the poorest communities. This pattern holds true across the globe, both in the modern world and in the past. It's not merely a question of access to a good doctor (though that certainly helps): structural inequality – the long-term establishment of differences between different groups within a society to that society's resources – kills. The inequality referred to here determines not only how much you eat, but how varied that food is, and of what quality; how tall or strong you grow; your resilience to things like infection and disease; how hard you have to labour (or how dangerously); and even the health of future children. The long-term effects of early-life experiences form the cornerstone of the Barker hypothesis, which holds that early-life health affects later-life disease.

The specific relationship that epidemiologist David Barker specified was between sub-optimal birth weight and foetal growth and the incidence of things like heart disease. Despite his theory falling down somewhat in that particular case, his conceptualisation of the effects of early life on the adult body has been retained, particularly in physical anthropology, as the 'thrifty phenotype' theory of disease and growth. The most straightforward example of what this theory predicts can be seen in the example of the Dutch. The Dutch are fantastically tall, the tallest in Europe, and certainly the tallest in the apartment I used to share in Santa Monica that was two-thirds occupied by a brother and sister of Dutch extraction, both of whom very much towered over me in their natural blond-coiffed height.* But in the aftermath of the Second World War, the Dutch were the *smallest* people in Europe. The 'thrifty phenotype' explanation for the obvious fact of my roommates' height against the history books comes from the Dutch 'Hunger Winter', the period of starvation under Nazi occupation and blockade that killed some 10 per cent of the Dutch population in 1944–5. When Amsterdam was liberated, there were literally people starving to death on the streets; children had been sent out into the countryside in the hope that they would survive the famine conditions of the cities. The theory goes that children born to parents who survived that extreme hunger were primed to survive on less, and they passed that ability down to their children. Those children took that ability to live on less and turned it into an ability to grow more, the 'thriftiness' of

---

* And of course, the *un*natural extra few inches the brother's pompadour gave him.

their genes finding an expression in their subsequent height. Unfortunately for science, the real story is much more complicated, with plenty of contradicting evidence, and in the end, we don't actually have all the answers for why Tiffany and John got so tall. It's also not very clear how I would recognise my roommates' potential exemplification of the epigenetic effects of heritable 'thriftiness' – they would have to be buried alongside several generations of their family, with very clear indications of who begat who, to even start to unravel that mystery. Happily, their annual family Christmas cookie party has not yet culminated in the sort of burial we'd need to really argue for visible trans-generation traits. But the Barker hypothesis retains potential, if not the hard evidence we require.

How can we identify the biological impact of inequality? In modern contexts, there are clear markers of nutritional deficiencies that can be traced directly to insufficient access to food resources. Marasmus is a polite term for a horrific condition: dietary deficiencies that can be translated more directly as starvation. Kwashiorkor is a similar childhood wasting syndrome caused specifically by lack of protein, and marked by skinny limbs and bloated bellies, hands and feet. The name for the condition itself comes from a Ghanaian language, which identifies the disease as that of the 'deposed' child: the weanling pushed aside at the birth of a new sibling. For communities on the very edges of food security, starvation is a real risk. According to the World Health Organization, malnutrition kills 3.1 million children under the age of five a year. For those lucky enough to survive food scarcity, starvation can still leave lingering effects in the form of stunted growth and later health problems. The first two years of life are critical in establishing normal growth patterns and achieving normal adult height. Subsequent interventions may reverse the course of malnutrition but may never recover lost stature, although humans, in their remarkable plasticity, do possess some capacity for 'catch-up growth', particularly in the two-to-four age range.

We know quite a bit about the danger of malnutrition from studies in the present and recent past. One of the most thorough studies comes from Australia, where indigenous groups were systematically pushed to the margins of survival by meticulously bureaucratic colonial authorities. Having pushed native groups out of their territories in favour of livestock and crops, and attempted to enforce 'civilising' structures of law and economics that securely demoted native Australians to the bottom tiers of society, the colonists proceeded to measure the long-term consequences of poverty and social breakdown. Aboriginal children, it was noted, suffered greatly from growth faltering in the first year of life. These children later 'caught up' some growth, largely in the next few years, but they would carry the scars of their early lives forever – in their height and in their teeth.

Teeth, as discussed in the first two chapters, work very well as a sort of fossil record of growth. Alongside reduced stature and signs of the infections to which malnourished individuals are more prone, it forms a sort of triumvirate of skeletal evidence for an under-resourced life. Studies of ancient bodies can reveal similar reversals of fortune. We have spoken about the decrease in height and muscle mass observed with the transition to agriculture, and the increased number of lines on teeth that signal some disruption in the life of the child. The third line of evidence that bioarchaeologists can use to investigate inequality comes from signs that an individual did not get the variety and quality of nutrients needed to sustain the body. Despite many popular claims to the contrary,* humans do not thrive if they follow a diet lacking in leafy greens, fresh fruit or other non-beige foods containing the vitamins and minerals that our bodies continually use.

* Until shortly after university, I remained convinced a diet of Red Bull, Marlboro Lights, pasta, butter and cheese was more than enough to sustain life.

There are a number of nutritional deficiencies that cause changes in the skeleton, which bioarchaeologists can later identify. The lack of specific vitamins can lead to very specific consequences for the skeleton, and, grouped under the heading of metabolic disease, these conditions can be used to reconstruct the holes in the menu offerings of archaeological lives. Bioarchaeologists Megan Brickley and Rachel Ives have more or less written the book on these,* and defined metabolic diseases as those that specifically disrupt the normal metabolic processes of bone: they interfere with the growth, mineralisation and resorption of bone. The complicated pathways by which our bones signal for change determine whether bone should be taken away or added to, and also decide where and when to store or tax bone mineral reserves, and they deserve careful analysis. The lazy assumption of simplistic mechanisms behind bone disease has caused countless misdiagnoses and considerable misunderstanding. However, I'm going to let you off with that warning, because really, what we're interested in is a) which metabolic diseases leave clues behind in the skeleton, and b) what those clues can tell us about the haves and have-nots of the past.

How do you end up with a nutritional deficiency?† An early and easily discountable argument is that some of the nutritional deficiencies we encounter were solved by the arrival of the Neolithic – having grain to hand gave us the supplies to survive harsh winters where we would have otherwise perished. There are quite literally tens of thousands of non-farming First Nations people like the Inuit who can set the record straight on that one. However, many more arguments blame Neolithic diet choices, without much thought for equality of access to resources, for the invention of nutritional deficiencies. We have heard all about the Palaeo Lifestyle Thing in Chapter 2, where nutrition comes wrapped in shiny foil in packs of 12. Assuming, just on balance,

* And continue to do so; I look forward to seeing the updated version that was under way at the time of writing.

† Do a PhD.

that hunter-gatherers in the past for some reason failed to invent energy bars, protein shakes and the other 'palaeo' supplements available for purchase in the modern world, what do we know about the risks of malnutrition in the 'egalitarian' past?

The argument we will examine here is that the transition to agriculture was a major risk factor for not just malnutrition in general, but for a number of conditions that are caused by lack of variety in the diet. Agriculture, this argument goes, produces a lot of food, but it's very same-y. The sports medicine guru Loren Cordain, whose book *The Paleo Diet* is at least partially responsible for the palaeo-diet fad, has written extensively on how the transition to agriculture was responsible for stripping our diets of nutrients by encouraging consumption of cereals. He argues that cereals are poor in nutrition, and humans only got the idea of foolishly eating grass seeds 10,000 years ago.

This is patently ridiculous. As covered in Chapter 2, we have eaten cereals for a *long* time. Recent work in Mozambique puts sorghum at least back to the Middle Stone Age – about 100,000 years ago. Also, it's hard to think of a more chicken-and-egg argument then one that says we only started eating cereals in the Neolithic – the corollary is that we only ate cereals *after* we'd decided to deliberately grow them.* The proof is in the (rice) pudding. Anyway, cereals work. They build more humans, which is pretty much the point. I do follow the logic that, if cheap calories are available nearby, folks may not go looking for the variety they once ate.† But is it true? Does the Neolithic encourage us to forget the vitamins and stay home and carbo-load instead? Or is something more insidious – like rising inequality in access to resources – to blame for a blander diet?

---

* I find it highly unlikely the Neolithic Revolution occurred because people collectively decided wheat flowers were pretty enough to cultivate. They are called spikelets for a *reason*.

† Ask literally anyone.

This is where bioarchaeology comes in. Nutritional deficiencies mess with metabolism, and where they mess with bone metabolism, the effects are discernible on the skeleton. In a few cases, there are specific vitamins – pretty much A through D – that have specific effects on bone. Vitamin A is a strange one: besides playing a role in the better-understood bone metabolism of vitamin D, too much of it can encourage bone loss. It's also possible to overdose on vitamin A, which is why you should never, ever eat polar bear liver, even if you're an adventurous European starving to death while exploring the Arctic.* Vitamin B comes in a variety of forms; we will specifically cover the case of a deficiency of vitamin B_{12} and B_9 (folic acid) here. Vitamin C, as any juice ad will tell you, is very important to health, and humans (along with other primates and, for some reason, bats and guinea pigs) don't make enough on their own to survive – we have to supplement what we can make with what we eat. Vitamin D is even more complicated, for similar reasons: mammals take in the vitamin D critical to growing bone through a combination of dietary sources and sunlight. If either of those is deficient, the bones can be severely affected.

Starting with the overabundance of vitamin A, it's hard to see how the Neolithic could mess that up for our species. As the problematic aspect is an excess of the vitamin, this is probably a special case; the major source of vitamin A poisoning is ingesting concentrated amounts that are stored in the livers of the top-of-the-food-chain predators like bears.† In fact, the earliest known case of hypervitaminosis A precedes even our species – it comes from the archaeological site of Koobi Fora 1.5 million years ago. The *Homo erectus* fossil fetchingly known as KNM-ER 1808 is argued to show characteristic signs of overdoing it on the liver front: on top of the normal bone

---

* Livers from animals at the top of the food chain have too much vitamin A; several centuries of explorers learned that the hard way.

† And palaeo-style diet adherents. Lucky for Hannibal Lecter, he preceded the popularity of the Atkins Diet.

structure, over-enthusiastic cells had started to lay down layers of coarse woven bone. KNM-ER 1808's tibias show a similar response to those of laboratory animals in hypervitaminosis trials, with the 'sausage casing' periosteum layer around the bone thickening into calcified bone and new bone forming, but with no changes to the inside of the bone cavity or to the skull. Before the age of vitamin supplements and synthetic vitamin A, it's hard to imagine a scenario where farming would lead to an increase in cases of poisoning-by-carnivore-liver, except in extreme conditions like ill-advised polar exploration.

Moving down the alphabet, we come to the case of vitamin(s) B. The specific skeletal evidence for vitamin B deficiencies we are going to discuss is actually an anaemia. Anaemias are a much broader category of problem, concerned with the number or amount of haemoglobin in red blood cells. They can be caused by a variety of nutritional deficiencies, or even straightforward blood loss. The body responds to anaemia with a series of staged interventions, one of which is to expand its capacity to make more red blood cells by making more bone marrow. The bones of the skulls are a prime location to see this in action – the inner table of the bones of the skull vault, the diploë, expand, and the smooth, hard surface of the outer table is lost. This creates a very distinct lesion called porotic hyperostosis, best imagined as a scattering of tiny pinprick-sized holes on the surface of bone, and it particularly affects the bones of the skull. For quite a long time, bioarchaeologists attributed the lesions found on the bones of the cranial vault to iron deficiency, citing things like menstrual bleeding and low social status (with concomitant lack of access to iron-rich meat) as factors in its appearance. After decades of argument, the late physical anthropologist Phil Walker and colleagues have made a convincing case that there are specific nutritional deficiencies that can cause this kind of skeletal response: lack of B_{12} and/or B_9. They point to a complex cycle of interactions from overall nutritional deficiency, caused by combinations of different deficiencies due to diet or even to parasite load. Walker's work suggests that famine conditions, particularly lack of meat, could easily produce B_{12} deficiencies, especially in infants who are breastfed milk already lacking in the vitamin. Coupled with the fact that growing children produce red blood cells in different places than adults, he and his team argue that this could be why typical lesions occur on the vaults of the skull and the upper part of the eye sockets.

The problem with children's eye sockets* is that they are better vascularised than adults, and thus susceptible to the sort of marrow expansion mentioned for anaemias. The lesions that form on the underside of the top part of the eye socket are called cribra orbitalia and loom large in bioarchaeology. There are a couple of ways to remodel the tops of your eye sockets, but the key to achieving cribra orbitalia associated with anaemia is in that marrow expansion. These changes are not restricted to vitamin B deficiencies, however, and may result from the interaction of several causes.

The lack of one vitamin may easily accompany the lack of another, which brings us to the evidence of vitamin C deficiency. Through association with the popular figure of the pirate, most people will be aware of the disease of scurvy. Scurvy is a syndrome caused by lack of vitamin C that causes hair to curl, gums to turn purple and peel back, teeth to fall out and bone to weaken. With insufficient vitamin C, no new bone is laid down and fractures can occur in the still-forming ends of weakened children's bones. As with the other vitamin deficiencies, the pinprick-like holes (porosity) are frequently observed on the bones of the skull and in other locations where pooling blood and inflammation have stimulated a bony response – including the eye sockets. Trauma to the eye or the weakening of the surrounding blood vessels due to the inflammation of scurvy can cause

---

* Aside from the fact that they are disproportionately large for their faces, giving human babies the eye-to-face ratio of the little grey aliens of popular imagination. Apparently, we find this cute in babies, dogs, cats and other animals, but put it in a spaceship and it's a sci-fi horror flick.

blood to pool, provoking a chain reaction that brings on the characteristic bony changes of cribra orbitalia.

Lastly, we have to consider the effects of vitamin D* deficiency. The pioneering work of the medical researchers Mellanby (Lady May and Sir Edward) in the 1920s brought to light the critical interplay of diet and sunlight in causing the devastating condition of rickets. In a critical – if cruel – experiment, beagles were systematically fed on diets designed to give them rickets (oatmeal) and kept indoors, away from the sun. The dogs promptly developed the expected bony changes, but were soon cured with a few spoonfuls of cod liver oil. Rickets is a condition that has been well known in medical history, and the characteristics of the disease are easily identifiable in the skeleton. The effects of vitamin D deficiency are linked to its role in calcium metabolism, and as every milk ad ever has averred, you need calcium to build strong bones. With no calcium coming into the growing ends of bones, they bend into strange shapes that are easy to deform.

With the childhood vitamin D deficiency, in addition to the scatters of porosity on bones like the skull, the growth plates at the end of long bones are the sites of major damage. With a lack of mineralisation, the growing end of the bone loses the neat lines of orderly growth and can flare out like a tassel. The diagnostic changes in the skeleton associated with the condition are largely based on the application of weight to bones that just aren't up to the job. Adding to this are changes at the junction of cartilage and bone, where profuse growth causes typical knobby joints. These are seen, for instance, at the junction of the ribs, in the changes in weight-bearing bones resulting in a characteristic bowing of the legs, in the flattening of the pelvis, and in changes to the ribs and sternum resulting in a 'pigeon chest' appearance. If bone mineralisation begins to pick up again, it will mineralise the

* I will flag here that vitamin D is not really a vitamin, it's a sort of proto-hormone, which is why its mechanisms are so ridiculously complicated that we still don't *quite* understand how it works.

now deformed bones, leading to lifelong skeletal changes; and these are what we can identify from archaeological human remains. Rickets and osteomalacia are two terms for the same disease, but the former occurs in growing bone and therefore only in childhood. Osteomalacia develops from exactly the same condition, but without cartilaginous growth plates to traumatise, is restricted to bone failing to mineralise and a reduction in bone mineral content; it's a lot harder to spot in the ground.

Coming back to the assertions of the man who would have you buy his diet book: Cordain has suggested that the agricultural revolution's limiting effects on our food choices are responsible for nutritional deficiencies. Admittedly, it's hard to argue for convincing cases of metabolic disease prior to the Neolithic. Scurvy, it turns out, predates pirates by a considerable margin,* but there are no certain cases that predate the Neolithic in the archaeological record. Conditions that seem very much like scurvy have been recorded by historians as far back as recording was done, but again these are limited to urban societies. Mesopotamian clay tablets describe a disease called *bu'sanu*, 'stinking disease', which many researchers have related to scurvy.† Archaeologists excavating a cemetery of the Predynastic Naqada culture at the site of Nag el-Qarmila found the remains of a one-year-old child with a number of porotic lesions and proliferative new bone growth, which fit the diagnostic criteria set up by the famous palaeopathologist Don Ortner. This brings the archaeological evidence of scurvy back to a time of villages rather than cities, though there remains the issue of what exactly was going on politically in the upper Nile Delta in

* Though this depends on how you view the eclectic commercial maritime tactics of the ancient world.

† It actually applies to any disease that gives you a stinky mouth; if you ever want to have fun at the dentist, try explaining that you follow ancient Mesopotamian dental theory and believe that tooth pain is caused by the toothworm, daughter of the goddess Gula, who must be ritually invoked to ensure satisfactory clinical outcomes.

the fourth millennium BC. Evidence from the urban cultures of Mesopotamia, Egypt and Greece fill in sketchy outlines of the rare presence of scurvy in the Bronze Age; in one case, an unfortunate Canaanite child from Israel seems to have undergone trepanation* to treat a scorbutic condition and died as a result. It's not just seen in urban sites, however: Simon Mays has reported an example from Bronze Age England. The twin stigmata of anaemic deficiencies are also seen in the Neolithic – there are reports from Neolithic samples in Greece and Japan.

An even more unfortunate child from the Chalcolithic period in the Atapuerca region of Spain is reported to have potentially had scorbutic legions *and* evidence of limb bowing related to rickets, though this is a slightly tenuous diagnosis. A recent re-dating of a century-old skeleton has suggested that one of the oldest probable cases of rickets is in a young woman from the Neolithic period at the site of Balevullin in Scotland. However, by and large, rickets is more often considered a disease of urban living. Modern medical reports clearly associate city living with a lack of access to sunshine, particularly in the industrial era. Indoor work, which will be covered in more detail in Chapter 12, seems to be as catastrophic for us as it was for the Mellanbys' beagles. The numbers of cases found in the archaeological record do not rise above the anecdotal until there are fully urban cities, with their endless possibilities for bad diets, pollution clouding the air and even buildings pressed so close together that they block out the sun.†

The case of the Glaswegian obstetric crisis of the early twentieth century is a clear example of this. The industrial city of Glasgow sits far north of the equator, and between the pollution of nineteenth- and twentieth-century industry and the traditional Scottish sunshine, very limited opportunity was provided for producing vitamin D. The gross poverty of

---

* With a scraping tool. And no anaesthetic.

† This would be the 'historic charm' so many cities now derive tourism income from.

the early industrialised city further accentuated the problem, and rickets was the exceedingly common result. While some deformity might be socially unsightly, for many young women it was a death sentence. The flattened pelvis of the rachitic skeleton could not accommodate natural birth, and maternal mortality skyrocketed. It took a revolution in surgical practice to save them, an impetus that would never have existed without the constant threat of rickets. On 10 April 1888, pioneering surgeon Murdoch Cameron performed the world's first known Caesarean section under anaesthetic; and while the women of Glasgow no longer suffer as their ancestors did, the rest of the world is undoubtedly grateful for the fruits of their labour.

Access to resources means access to adequate nutrition. Certainly the modern world has numerous examples of nutritional deficiencies relating to poverty and the ability to obtain a variety of provisions. Sometimes we have overwhelming evidence of nutritional deficiency without the archaeology to back it up. One of the interesting (and I think under-explored) areas of bioarchaeology is in the microscopic study of metabolic bone disease. Certainly a topic of considerable importance in clinical literature, it has been a slow build-up for these approaches in archaeological practice. Bioarchaeologists often do not have the necessary equipment or funding to carry out the incredibly time-consuming work of creating histological thin sections, though this is starting to change.* However, under the microscope, a whole new world of bony changes becomes obvious; freed from the restrictions of trying to work out 'what disease' based on where in the skeleton, we can look at the specific mechanisms behind disease. The specific actors in each scene of a bone's construction are easily identifiable under a microscope, and when they don't hit their marks at the start of each act, we can see evidence of metabolic disease. The curious case of

* Histological thin sections are how people make slides out of slices of bones and teeth. It's enormous fun, if you like repetitive grinding and polishing activities. Therapeutically relaxing, really.

death-by-corn gives us an example of a disease we know from modern medical studies affects bone, was definitely present in the past and may be an important clue to periods of major social upheaval where adequate food was scarce.

One might not suspect that, of all the American exports of the last few centuries, corn might be one of the most dangerous. As the pedants among us will be aware, 'corn' in this context is specifically maize, the native American grain we discussed in Chapter 2.* Fast- and close-growing corn was a cheap menu option, and as soon as the European colonists realised this, they started sending it home. Maize and maize flour were exported all around the world. Unexpectedly, all around the world people fell ill. Those who were stuck with maize because poverty offered little choice in terms of their cereal intake would experience the triple-Ds of dermatitis, diarrhoea and dementia; these would be followed after a few years by the Big D itself. The Italians called this 'sour skin' disease, *pelle agra*; the modern term is pellagra. It took until the twentieth century for some bright spark to notice that the longest-term consumers of maize flour – continental Americans – did not seem to suffer from pellagra. Maize, it turns out, grows fast but grows mean – the critical micronutrient niacin that humans get from their cereal grains has to be coaxed out of corn flour by chemical means. In Mexico, maize flour is soaked in lime juice,† which releases the niacin into a form human stomachs can absorb. The exports never went through this process, so for those whose choices were limited by desperation and poverty, pellagra was a real danger. Niacin deficiency is thought to have played a part in famines affecting the nineteenth-century Irish; impoverished, maize-dependent communities

---

* 'Corn' more properly means any sort of grain; Americans have taken the application of the word to their native staple grain and sort of 'World Series of Baseball'-d it.

† As far as I can tell, there is a general theory of Mexican cuisine that holds there is nothing – and I mean nothing – that can't be improved with the addition of more lime.

in the twentieth century; and more recently, refugees and displaced persons in the twentieth and twenty-first centuries. Pellagra has been identified through a combination of macro- and microscopic observation on twentieth-century remains; perhaps when bioarchaeology budgets are a bit better, we will start to see evidence of this kind of dietary deficiency in the archaeological record as well.

What we have seen from the nutritional evidence is that, while isolated cases of vitamin deficiencies go well back into the past, we cannot say that biological food inequality really took off until the invention of the urban world. With the rise of cities, we can see a divorce between the food producers and the food procurers. The extra step of acquiring food through the means of exchange seems to throw up roadblocks for a substantial number of individuals. In rural contexts like the American South in the plantation period, we know that even seriously malnourished individuals were able to supplement their meagre rations with wild-caught food; excavations of slave quarters in North America, for instance, have uncovered caches of fishing hooks and raccoon bones. Furthermore, the dense populations and unsanitary conditions in many cities encourage infectious disease and parasites that can interrupt normal metabolic function. As with the evidence for interrupted growth trapped in teeth, a pattern that starts in the villages of the Neolithic becomes full-blown in the context of urban life.

So, we must return to our original question. What is a city, and what does it do to us? The key difference, when we begin to talk about cities, and rather more particularly the effect of cities on our species, is the numbers of people involved, how long they stay put … and the number of bureaucrats required to organise their social and spiritual wellbeing. Temple management requires at least two types of people, in a social world that for several hundred thousand years* had been largely egalitarian, and if this is a year-round temple, with year-round functions, you will require a full-time bureaucracy.

* Or several million, depending on your definition of 'human'.

So perhaps the cathedral does mark the city. But as cities come together, we start to see that not all parts of the new urban world are created equal. Tortoise people or temple people and hoi polloi go into the ground with markers of their in-life status: jewellery, pottery, stone vessels, metal weapons and pins. Entire subsections of the city have different characters; fancy tombs stand apart from those of the huddled masses, and when we look at the stories written into those bones, the ones from fancy tombs tell a different story than the ones from workhouses and paupers' graves. And yet, for all that, cities flourish; the story of urbanism is that we kept building, and the people kept coming. The success of the city is the same as the success of the Neolithic; it's winning at a numbers game. However, even that prize is not without its costs.

CHAPTER SIX

Oops Upside Your Head

Cities are focal points for inequality in the past. The ranked differences among their inhabitants – in terms of the work they do, the resources they can acquire and the sheer scale at which all these are carried out – is what fundamentally separates cities from the Neolithic world of villages and towns. Cities are also associated, if we paint the past in very broad brush strokes, with the intensification of new technologies. Cities are usually associated with the dawn of the metal ages in Europe, Western Asia and Africa,* a complicated interplay of resource extraction and craft specialisation. One of the earliest exploitations of metal seems to have kicked off in the mountains of what is now Serbia among the dense villages of the Vinča culture 7,000 years ago; however, it's not until the dawn of the urban age some 2,000 years later that metals become the lifeblood of the world's first arterial trade networks. What was a symbolic and functional invention becomes, in the presence of so many people and so much hierarchy, a commodity. I have seen with my own eyes the delicate Anatolian animals, cast in unyielding bronze, cached in the tombs of the people who straddled the trade routes of the early Mesopotamian cities.† Of course, one of the other forms of metal from that very same site is a spearhead, and when we look at metal as it circulates through these early empires, it's as much as a weapon as a decorative art. Weapons are older than metal, and violence is older than cities, but these next few chapters take a closer look at the possibility

---

* Metal doesn't equal cities any more than tortoise shells equal inequality; it's just one more type of thing that you can show off, trade and complicate your life with.

† And I've listened to the site bronze specialist describe my own pitiful findings of bronze clothing pins as 'not very good'.

that cities, with all their weapons and all their people, have intensified one of our oldest traits: violence.

What can we say about violence in the past? And what's more, does anything really need to be said? Anyone who catches the news will be fully aware that our species is doing its level best to live up to Hobbes' description of 'nasty and brutish'. We are a violent species in a violent clade, and the evidence for this can be extracted from the skeletal remains we have left in our bloody wake. Readers who have ever encountered, interacted with or been small children will no doubt be intimately acquainted with our seemingly inherent hair-pulling, pushing, biting and scratching nature. However, what do we really know about our violent tendencies? Are we evolutionarily engineered towards it, with only a thin veneer of civilisation preventing us from going full *Lord of the Flies*? Or is it civilisation itself, with its vainglorious wars of conquest, its capital punishments and its relentless competition, which moves the hand that holds the sword?

This is a rather big question, and will actually occupy the next three chapters. To try to get a handle on things, I've broken the discussion into three categories: interpersonal violence, structural violence and community-scale violence – the latter better known to poets and pedants as war. This chapter will introduce the techniques that allow us to reconstruct evidence of violence from the skeletons of the past. These techniques – forensic, archaeological and bioarchaeological – are the same ones we use to identify structural violence and warfare in the archaeological record. This chapter, however, will set the stage for understanding the origins and patterns of conflict through evidence of violence at its most basic level: one-on-one, me against you.

The antiquity of people smacking each other upside the head must be presumed to go far, far beyond the invention of urban life, so why do we need to think about it here? Much like the impacts of disease, which have always been with our species, urban life acts as a sort of hothouse for the social, economic, cultural and psychological factors behind interpersonal violence: human competition, for status, for access to resources or even for sport.* We know that violence did not begin in cities. Life on earth is violent, whether you're a cognitive marvel of primate evolution or an unsuspecting, happy-go-lucky spider that just happened to be in the wrong place at the wrong time. Life for our little clade of humans and their ancestors has been very violent indeed, and we have the bones to prove it. For example, we have an approximately 430,000-year-old skull from Spain that seems to be pretty unambiguous proof that we were bad before we were even us. The north of Spain is a karstic region that is absolutely littered with caves. Buried deep within are the last traces of some of our long, long lost relatives, and at a site in the Atapuerca Mountains near modern-day Burgos, archaeologists uncovered an even larger surprise.

The site of Sima de los Huesos translates as 'the pit of bones', which is a very apt description. The bones in question are from hominid 'ancestors' dating back almost half a million years, and they have yielded fascinating insights into the world our ancestors inhabited. Around 28 human(ish) individuals were found in the pit, alongside more cave bears than you could shake a stick at. Cranium 17, one of the skulls found at Sima, has holes in it. This is less surprising than you might think – after having probably fallen into their position from elsewhere, hundreds of thousands of years in a cave largely occupied by giant extinct bears might be imagined to have had knock-on effects on any bones underfoot. However, a recent publication from the Sima de los Huesos research team suggests that the holes in the adult male skull are more than just accidental damage. They argue that the two holes in the skull are clearly evidence of repeated blows with the same weapon: signs of lethal intent. The blows seem to have occurred around the time of death, which for the authors of

* Though I would argue that sports like gladiatoral combat actually count as structural violence, considering the violence done to them was by the will of the society they lived in; it's a fine point, but tridents have three and they don't look like something you'd chose to live by if you could avoid it.

the paper makes Cranium 17 the first *known* man to be murdered in all of human history.* Many of the other bodies found with Cranium 17 also show signs of injury, so perhaps he wasn't alone in his violent death. While this has been heralded as the very first sign of interpersonal violence seen in our entire human history, physical anthropologist Tim White has made a case for cannibalism, if not murder, in one of our early ancestors, an early member of the genus *Homo* found with stone-tool cut marks in the South African site of Sterkfontein dating back 1.5 to 1.8 million years ago. However, as discussed further below, the difficulty in identifying these 'firsts' from the archaeological record is that it can get very difficult to tell a killing blow from a bit of damage to the recently deceased.

I first became enamoured of the study of the human past on a long-ago course in physical anthropology, taken on a whim at a local community college in Southern California.† After that, I was hooked; I changed my major to archaeology and went off to UCLA to learn more. At various points in my undergraduate studies, the subject turned to the Chumash, a widely spread indigenous group that had occupied a stretch of coastal California from Los Angeles up past San Luis Obispo. Theirs was a fabulously rich and complex culture encompassing the manufacture of prestige tradeable shell beads (essentially, money) and a (small 'b') byzantine network of social alliances that would put the Borgias to shame,‡ before the Chumash people were eventually decimated and enslaved by Spanish missionaries. Despite the brutal nature of repression imposed by the Spanish and the Church, the Chumash culture survived the Mission period, and they take an active role in researching

* Unless you count the example in Chapter 8.

† Presumably, underwater basket-weaving was full. Thank you Orange Coast College.

‡ Explained with commendable brevity by the local Chumash representative on my very first field school in Malibu, California, a remarkable woman with long, long hair and a voice like Janis Joplin, in the following terms: 'We f*cked *everyone*, man.'

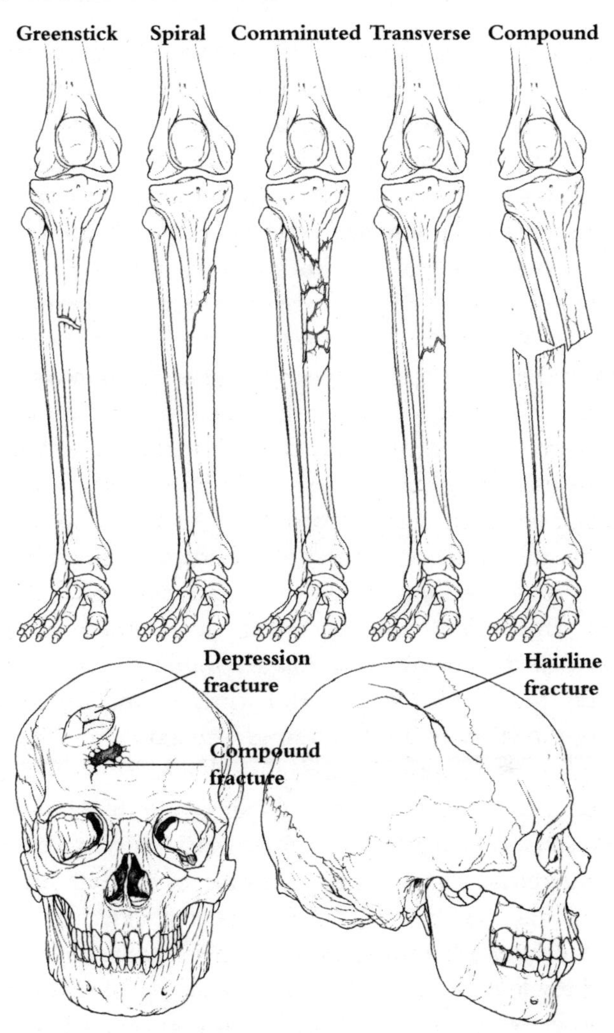

Figure 4 *Illustration of basic fracture types as applicable to long bones and the skull.*

and sharing their long and fascinating history. The physical anthropologist Philip Walker had an excellent relationship with the Chumash individuals responsible for their cultural heritage, and devoted much of his impressive career to studying the remains of their ancestors.

One of the most interesting things his research uncovered was that several Chumash skulls possessed a notable lumpiness. This was not some sort of genetic deformity, but rather a series of welts right on the skulls themselves. Walker discovered that the Chumash skulls showed high numbers of depressed cranial fractures – skeletal evidence of trauma to the skull that had been sustained during life. The lumpiness arose from where fractures to the bones of the cranial vault (the dome of the head, more or less) had healed over unevenly. Of course, there are many ways to smack your skull; in the modern world, low-lying lintels and unexpected lighting fixtures are the blight of the human with the audacity to be of over-average height. However, the impressions in the Chumash skulls were regularly circular, or at least approaching circular. This begins to suggest that the prehistoric Californians were less clumsy than confrontational.

The Chumash were a large group, with marked internal hierarchies; this must have thrown up sufficient occasions for inter-group tensions to reach boiling point. Aside from reading the story of a Chumash girl abandoned for years in the California Channel Islands around the time of Spanish incursion,* I knew very little about Chumash culture, and was reasonably delighted to follow Walker's suggestion that they had perhaps taken up a rather unique method of conflict avoidance known from other Californian cultures. Further north, another group of coastal Native Americans from the Monterrey area of California went to the trouble of establishing a very ritualised form of grievance procedure, witnessed by a somewhat judgemental military attaché of the colonising Spanish forces, one Pedro Fages:

> If two of the natives quarrel with each other, they stand body to body, giving each other blows as best they can,

* *Island of the Blue Dolphins*; merely one part of the California public school system's attempt to atone for historical genocides – other required reading included *Sadako and the Thousand Paper Cranes*, the story of a girl in the aftermath of the atom bomb attack on Hiroshima. Fourth grade was a sobering year.

> using what might be called spatulas of bone, which they always carry for the purpose of scraping off their perspiration while in the bath and during the fatigue of their marches. But as soon as blood is drawn from either of the combatants, however little he may shed, the quarrel is forthwith stopped, and they become reconciled as friends even when re-dress of the greatest injury is sought.

Archaeologists have indeed found evidence of 'spatula-like' bone tools, interpreted as being similar to Roman strigils and used to keep the body clean by scraping off sweat. Bioarchaeological analysis of the skulls uncovered in excavations of Chumash remains can identify signs of blunt force cranial trauma either visually, spotting literal dents in the cranium, or through radiography, using X-rays or CT scans to spot areas of dense bone that are the hallmarks of healed trauma. These are exactly the lesions you might expect from the sort of spatula-spat Fages had described. The case that captured my attention, all those years ago, was that the skull of one elderly Chumash woman showed evidence of *nine* separate healed lesions; one would like to imagine this a testament to her ability to argue a point, or at the very least, her tenacity in doing so.

Of course, it's not guaranteed that the lumps and bumps of the Chumash skulls were caused by anything so orderly as ritual conflict settlement. Forensics and modern hospital records demonstrate the harrowing statistics behind blows to the head in the modern day; road accidents in particular are the leading cause of traumatic injuries, distantly followed by falls and assaults.* A recent study in India estimated that about 60 per cent of traumatic brain injuries were due to the modern invention of the car accident, with a further 20 per cent explained by falls and around 10 per cent explained by violence. These numbers vary considerably around the world and among different social groups within even the same

---

* This is exceptionally interesting literature to review while preparing for your motorcycle licence.

countries; a report calculating statistics for the latter half of the twentieth century found that the incidence of head injuries from violence was 40 per cent for inner city Chicago and 4 per cent for a more bucolic suburb, with higher rates in men than in women. These are of course clinical reports, which must be considered in a different light to the evidence we can see on skeletons – we have no living witnesses to report that someone has fallen on their head, and no soft tissue lacerations to testify to injury.

Much of what we know about injury comes from modern medical studies, conducted on populations in the developed world who have access to hospitals of the sort that are stalked by epidemiologists. This works very well, up to a point – for example those road accidents, which are unlikely to account for the majority of archaeological cases. Both clinical study and forensic investigation contribute to what we know about broken bones and skull dents found on human remains from the distant past. Through clinical studies looking at population-wide trends, we know that certain types of fractures are more likely to be caused by specific activities, falls or impacts. For instance, a small study looking at the number of under-18s at a trauma clinic in New Jersey found that about 43 per cent of fractures treated were caused by interpersonal violence, a much higher rate than that commonly reported for a general population. Most of these occurred in teenage boys,* with the most common fractures being of the nose and the jaw. Cuts, scrapes, impact craters and perforations in bone can be traced to specific implements. Moreover, the prolific violence that our species is known for means that there is a wealth of forensic studies that can place the number, angle, depth and position of damage to the skeleton in context, telling us everything from the emotional state to the handedness of the attacker.

Here the work of Phil Walker again comes to the forefront. In a relatively ground-breaking review in 2001 (the same year

---

* Surprising no one.

I was driving to Malibu at 5.00 a.m.* every morning to complete my undergraduate fieldwork on a Chumash shell mound† site at Point Mugu State Park campground), he laid out a bioarchaeological theory for understanding the lumps and bumps on the Chumash skulls he had observed, and more besides. While the only type of violence we can interpret from the past, he admits, is limited to the physical violence that leaves scars on the bones themselves, there are basic principles that can allow us to reconstruct a blow-by-blow account of violent injury and death in the past from the skeleton. Specific types of fractures signal particular likely factors, and they can largely be categorised into how the bone broke and where it broke. Taking the latter first, fracture location is one of the main clues to the activity that caused the break, and these are usually grouped into cranial (head) or postcranial (the rest) in large epidemiological studies. Cranial fractures can either include the bones of the skull vault (the cranium, the bits without a face on) or the face itself, and they are one of the most common sites for fractures, now or in the past. Facial fractures by far and away most commonly occur in the nasal bones; then as now, a broken nose is likely to suggest some sort of contretemps.

The depressed cranial fractures that Walker observed on the Chumash are one of the most studied pathologies in bioarchaeology. Walking through the process of formation allows us to see how knocks to the head that the victim survived for some time can be differentiated easily from fatal

---

* That was the goal, anyway. One (male) member of my carpool discovered that if we hit the Malibu Starbucks at *exactly* the right time we'd end up in line behind Pamela Anderson, so … apologies to Professor Arnold for being late. A lot.

† For anyone unfamiliar with the archaeology of coastal California, a great deal of it is taken up with mounds of shells that can be metres deep. The one in Point Mugu was big enough, and contained only three kinds of shell for the most part, which is pretty much the most boring thing in the world if you're stuck with the job of sorting them (sorry, Professor Arnold).

blows. Anyone who has ever broken a bone and had it knit back together will be familiar with the idea that, if treated properly, a broken bone will heal. This can be contrasted with teeth, as discussed in Chapter 2, which, when chipped or broken, stay that way.* Bone constantly remodels throughout life, with specialist cells eating away at the insides while other specialist cells lay new bone down on the opposite side. This process is ongoing while an organism is alive; there isn't a bone in the adult human body that is really 'older' than about 11 years, because the remodelling process constantly replaces bone over time.

A smack on the head with sufficient force causes a cascade of reactions. This is blunt force trauma – a word bandied about by television hospital soap operas and hack novelists alike, so you can be sure it's a fairly simple concept. Blunt force trauma is the wound you get when you're hit (hard) by a non-pointy, non-projectile weapon. Colonel Mustard in the library with the candlestick – this is how we end up with skulls with depressed cranial fractures. The bone itself might be smashed in, either right the way through or, in less percussive cases, only denting the skull. The periosteum, the sensitive membrane around a bone that produces all the cells for building up or taking away more bone (and which I have in the past asked my students to imagine as a sort of sausage casing around the bone), signals for a host of resources to be trucked into the affected area, forming a bruise† that pools the material for repair. The broken bone tries to knit itself together with new bone, first forming a loose bolus of cartilage around the fracture site, before replacing this soft callus with woven-looking bone – a process that takes several months. If the bone is more or less in the right place, the callous is gradually resorbed until the bone looks pretty much the way it did before the break. If the bone has been dramatically shifted out of place, healing can result in dramatic lumps, the callus constantly laying down in exotic

---

* Not to mention extending the pain to your wallet.

† Fancy talk: haematoma.

new directions as it tries to reach the disparate ends of the bone. In extremity, entire false joints can be created if the broken bone keeps being pressed into service before the fracture ends can mesh back together. If the break extends through the skin (a 'comminuted' fracture), you add the excitement of infection and septicaemia to the mix. With cranial trauma, if the blow hasn't killed you, nor the build-up of pressure from all that extra repair material rushing to the site of the wound, the skull will heal over; but, as with many DIY repair jobs, the rebuilt area may not be quite on level with the rest of the structure.

The healing process is one of the ways that physical anthropologists can identify fractures, but if the healing processes all went swimmingly we wouldn't have any evidence at all. Luckily (for science), the techniques to reset fractures and immobilise limbs that we use today were not necessarily always employed in the past. It should be remembered, however, that the evidence of injuries to bone can disappear with time; what we see in the skeletons of the past is evidence of either poorly healed injuries or injuries that were still healing at the time of death. The mechanism of healing fractures allows us to establish a timeline for wounds we see in bones. A small, very dense lump around a bone with a discontinuity visible in X-ray might suggest a nearly healed fracture of several years' antiquity; a hole in a skull with tiny microscopic evidence of a few bone cells stirring around the very margins would suggest that healing was an abortive process interrupted within a matter of hours or days by death. The former is easily identified as an antemortem injury (occurring before death), while the latter might be considered a perimortem injury (occurring around the time of death). By contrast, a great whacking hole the size and shape of a mattock end found in a skull on an archaeological site in close proximity to a nervous and/or sheepish-looking archaeologist is likely to be a very postmortem injury, the result of 'taphonomy' – a fun Greek way of describing the things that the environment[*] does to the body after death. Forensic archaeology makes great use of taphonomy to determine timelines for body decomposition, insect invasion and burial practices like ritual postmortem dismemberment. The 'body farm' run by the University of North Carolina at Chapel Hill is perhaps the best known of these experiments, though considerable research is done in unexpected locations throughout the world.[†]

Skulls are also rather uniquely suited to puncturing wounds. It's for the same reason that depressed cranial fractures are so common: skulls contain a great deal of surface area and sufficiently integral parts that they are frequent targets for attack. While the medical literature attributes most depressed cranial fractures[‡] to blunt force trauma, evidence of sharp force trauma – which is very much what it sounds like – is also commonly observed on the skull. Sharp force is differentiated from blunt force mostly by the area of bone impacted, and in a forensic context usually implies wounds made by weapons. Chopping, cutting ('incising'), slashing and stabbing are the main modes of inflicting sharp force trauma, and, as any half-decent prison show will demonstrate, *anything* can be a weapon.[§] Differentiating between the two types of trauma may occasionally be difficult. An axe may well stave in quite a bit of skull, but is technically an edged weapon; blunt force with enough force can push through just as much cranial bone as an axe. Forensic anthropologists and bioarchaeologists must

---

[*] And/or careless archaeologists.

[†] Including one unassuming North London garden owned by my colleague Ros Wallduck, who has had to do a *lot* of explaining to the rubbish collectors. And the local butcher. And her husband.

[‡] This does of course include traumas like those caused by modern pursuits, such as crashing fast-moving cars into stationary objects and hurtling down snowy hills directly into stationary trees wearing sticks on your feet.

[§] Even spoons. Particularly spoons – see Alan Rickman's Sherriff of Nottingham in the unaccountably underrated 1991 classic *Robin Hood: Prince of Thieves.*

use whatever clues remain to piece together the method of fracture. Where thin little breakage lines snake across the surface of the bone, like fissures in dried mud, it's possible to trace these lines back to their source to try to gauge the original direction and impact of the blow. Of course, it's far easier to reconstruct the causes of fracture if you can simply match the edge of the weapon used to the dent made in the skull, and even better if the weapon is still wedged in.

Getting back to the rest of the body, postcranial fractures can also be caused by accidents or either blunt or sharp force trauma, blunt force being more likely to cause a fracture. These fractures are broken down into their constituent bones or functional limbs. There are too many potential fractures to go into for practicality,* but there are a few basic types that come up again and again in forensic and medical literature. Wrists and forearms are pretty key sites for damage. Another fracture that affects the forearm, and a common one in modern medical practice, is Colles' fracture, which affects the part of the radius that meets up with your wrist – if you put your hand down flat on a table, fingers pointing away, you can feel the lumpy end of the radius on the thumb side of the hand. Like most wrist fractures, Colles' occurs when you fling your hand out to break a fall, and isn't necessarily diagnostic of violence. The wrist ends of both radius and ulna are frequent sites for this kind of fracture.

As an experiment, stand with your hands at your sides. Imagine that you are having a perfectly nice time standing around and minding your own business, when suddenly – an attack! Something is coming at you from in front and a bit above; perhaps it's a candlestick, perhaps it's a rain of frogs. Either way, if you instinctively brought up an arm to ward off the airborne amphibious menace, then you would understand the underlying mechanism of the 'parry' fracture. The parry fracture takes its name from the action of parrying off a blow

* Doctors very much seem to enjoy naming newly discovered ways of snapping bones after themselves.

or attack,[*] and affects the ulna, one of the two bones that make up your forearm. The ulna is the bone that forms the knobbly bit of your elbow; if your hands are flat down at your sides, it's the one in the back (the other is the radius). If you have one hand in the air, with the palm down, and someone is coming at you with a weapon, it's the ulna that gets it and, if the frog has sufficient velocity, the ulna that breaks.

Parry fractures are one of the most frequently cited archaeological indicators of violence. In his summary of evidence for forearm fractures relating to violence in the past, eminent physical anthropologist Clark Spencer Larsen notes that parry fractures are more likely to be related to interpersonal violence when they disproportionately affect one side of the body in a population. The over-representation of left forearm fractures might indicate that a right-handed person was defending against attack, or conversely an armed right hander might have their weapon in their dominant hand, leaving the left to bear the brunt of an opponent's strikes. Larsen notes injuries following this pattern from cultures as varied as groups of Australian Aborigines to native Hawaiians, but this is not a very representative sample of all the various cultures that have ever fought among themselves (or others) with weapons liable to break arms. As with all bioarchaeological evidence, more support is needed to clearly identify signs of violence. One potential additional indication of whether a fractured forearm is a clue to violence or to clumsiness is in the people it affects. If adult males make up most of the affected, then you might speculate that the injuries sustained relate to a social role they play. If one of those social roles involves clubs, shields or the aggressive deployment of both against other human beings who *also* have clubs and shields, you might well expect a higher number of parry fractures in adult males to be the bony evidence needed to reconstruct how those men lived. Similarly, if fractures are equally distributed among men and women, and it's clear that otherwise they occupy quite different social roles, the

---

[*] Admittedly, rarely involving frogs.

interpretation of fractures found might lean towards accidental injury.*

The rest of the postcranial fractures build up similar lines of evidence. Ribs can be fractured accidentally or violently; even coughing too hard can be enough to crack one. Sometimes it's the type of fracture, rather than the location, that gives away the violent origins. A majority of radiologists consider metaphyseal fractures, where the still-growing (and therefore less firmly attached) end-plates of children's bones are literally shaken loose, to be a damning indictment of child abuse. Stress fractures can result from habitual activities taken a step too far,† and commonly occur on the feet and lower legs of athletes, military cadets and keen Jazzercisers.

Fractures sustained at different times of life can have subtly different effects on the skeleton. The best known of these is the 'greenstick' fracture that occurs in kids, so called because the high plasticity of the juvenile skeleton leads to long bones shearing in a rotating direction; just imagine your six-year-old forearm being twisted and broken like a twig off a tree. My personal experience of fracture is just such an occasion. Having inherited a pair of roller skates of unquestionable style and coolness,‡ I was determined to skate everywhere. This included piles of loose dirt, which meant that I promptly went tumbling and landed heavily on the arm I threw out to catch myself. I managed to fracture something in my elbow, but my bones were still growing and nothing was knocked firmly out of place; the sausage casing of the periosteum was presumably intact, so my growing bones could knit together in more or less the right shape. If I'd repeated this exercise later in life, say with a borrowed bicycle

* Or a particularly gender-neutral pattern of violence.

† I can think of one particularly glamorous acquaintance whose years of dance training combined with a love of high heels have seen her accrue sufficient metatarsal fractures that she now has a wardrobe full of elevated flats just the right height to match a walking cast. Both sides have been worn.

‡ They looked like blue Adidas sneakers but with *wheels*.

and the steep hill on the north side of Finsbury Park in London, I probably would've managed a Colles' fracture.

Of course, as with almost all categories of skeletal evidence, we require multiple strands to make the case for violence. As bioarchaeologist Margaret Judd has pointed out, a stress fracture to the ulna caused by repetitive actions (like certain sports) can be indistinguishable from an authentic parry fracture. However, it's possible to overcome the inconclusive nature of the vast majority of archaeological evidence of violence.* A prime example is the co-occurrence of parry fractures with evidence of injuries to the head, which signals that there were blows being aimed in that direction and actively warded off. However, our recurring problem in the study of fracture patterns is the unfortunate lack of fractures identified.† Without all of these compounding lines of evidence, the life of the bioarchaeologist is rendered difficult indeed.

Not all fractures are the result of violence, of course. Hip fractures, for instance, are one of the most common injuries sustained by older females, but no one is particularly concerned that this is the result of a sustained campaign of violence against little old ladies. Hip fractures just happen to be one of the many painful consequences of osteoporosis, and it's older women who are most likely to have osteoporosis. It's also worth reiterating that a well-healed fracture, like my elbow, might completely resorb into a perfectly normal-looking bone with time. Even the telltale differences in bone density visible on an X-ray or CT scan might resolve themselves, leaving no one any the wiser.

So, we must move beyond the evidence of fractures. They are not the only clue we can find to violence in the past, and we can see further evidence of violence from sharp force trauma. The sharp edges of weapons (or tools, or ice picks)

---

* As will be emphasised further below, not all evidence is circumstantial. It is, for instance, very hard to argue with an ice pick to the head.

† For us, not the people of the past. Obviously.

can scrape tissue and even bone on their way across or through the human body. There are a few basic forensic principles that allow us to reconstruct some of these injuries, though it should be remembered that we know a lot less than we could about how these injuries form because it's considered bad form to go around stabbing people in slightly varied ways just to see what their bones look like afterwards.* On that same count, however, we know rather more than we would like, because there are a great number of people in this world who don't have the kind of moral constraints a university ethics board imposes, and the bodies they leave behind form the backbone of our understanding of traces of sharp force trauma. The basic concepts are actually very simple: weapons with a cutting edge meet bone and interact in a few circumscribed ways, because physics. A slashing attack that reaches bones will be more likely to leave a long, but not deep, cut mark on the affected bones. A stab, by contrast, might leave a very deep mark, or an inconsequential nick; most stabs are aimed (if they are aimed at all) at the soft, vital bits of the human body. An axe – well, it's fairly obvious what axes do. The more force behind an injury, the more difficult it is to tell what caused it, because there is likely to be much more damage to the bone, obscuring any initial punctures, scrapes or scratches on entry.

Even when it comes to dealing out death and violence, there is a 'correct tool' for every job. This is an adage that hangs on the human propensity for laziness just as much as it does our propensity for invention. The idea of using whatever is to hand, culturally or practically, applies to the act of violence just as it applies to mowing the lawn. Specific weapons leave their own specific wounds in the skeletons of the dead, and just as tool cultures might vary from group to group, the weapons of choice can also follow cultural lines. The earliest argued

* That said, doing it to corpses and dead pigs is considered a legitimate research exercise.

case of a proper weapon causing injury comes from the cave site of Skhul, in modern-day Israel, somewhere between 80,000 and 100,000 years ago. A spear was used to stab an anatomically modern human male, Skhul 9, in the left leg once or twice before sinking all the way into the pelvic cavity. The blow would have been fatal, and as archaeologist David Frayer has described in the separate case of a Mesolithic man with telltale stab marks on his vertebrae, unlikely to have been caused by having pretty much the worst Buster Keaton moment ever next to a pile of spears. However, what looks like murder to some looks like postmortem damage to others: physical anthropologist Erik Trinkaus espoused the position that the damage was caused after death, meaning that we aren't looking at a cold case from a time when we could pin it on the Neanderthal neighbours.

Phil Walker made particular note of these cultural leanings in his survey of violence: the British, for instance, sustain (and deal) a disproportionate number of the pint-glass-based injuries in the world.* In a similar vein, living in North America dramatically ups the odds of encountering a baseball bat in an unsporting way. In much the same way, the injuries weapons meted out in the past can vary from group to group and region to region. Occasionally, even the evidence of medical invention is a clue to medical necessity. Take for example the Inca, who ruled their empire with clubs and … cranial surgery. The extraordinarily high number of skulls found with holes deliberately drilled into them while the owners were still alive has been at varying times interpreted as ritual or spiritual, but a study of several hundred skulls found that many of these surgical interventions were associated with cranial trauma. The use of clubs as the primary weapon in Inca warfare and the subsequent blunt force trauma they

---

* Note to non-UK residents: This is called 'glassing', and involves some inexplicable deep-seated reflex to smash your glass on a table and stab people with the shards when they annoy you. If someone offers to demonstrate this to you, consider leaving. Quickly.

could cause may account for this remarkable trend. The antiquity of this traditional form of battle may also explain how Inca surgeons had gotten so good at drilling holes in heads by the 1400s that they achieved nearly 90 per cent survival rates.*

Sometimes the evidence of violent damage to the skeleton is more oblique and requires a circuitous logic to be recognised as such. Sometimes, it's just easy. Bones are not the only part of the human body that can be damaged. On occasion, the soft tissue of archaeological human remains is preserved sufficiently that we can identify marks of violence almost in the same way that a forensic investigator might address a recent homicide. Organic preservation is subject to all sorts of caveats. Some humidity is bad, but total immersion in peat is great; a bit of heat will speed up deterioration, but a lot of dry heat will lead to mummification. There are environments all over the planet that either speed up the destruction of soft tissue or miraculously preserve it: arid deserts, frozen tundra or even anaerobic peat bogs. While desert-dwelling mummies might be better known, it's actually in the very cold parts of the world that we see some of the most surprising finds preserved, up to and including a little 40,000-year-old baby mammoth from the hard permafrost of Siberia.†

Bodies do not appear out of deep freeze very often, but when they do, the results can be spectacularly surprising, a sign that the stories told by skeletons are perhaps far more circumspect than reality. The very particular case of Ötzi the Iceman is exactly such an occasion, and gives us reason to doubt our ability to 'find' violence in the archaeological record as easily as we sometimes seem to think we can. Ötzi is the name given to a man who died in the Italian Alps sometime around 3350 to 3100 BC, over 5,000 years ago. His body miraculously ended up in such a position that his remains were largely saved from the crushing weight of the glaciers, and he and his few possessions froze into the landscape

---

* One satisfied customer even returned for surgeries seven times.

† *e.g.* the adorable baby mammoth Lubya.

until an ice melt and some intrepid hikers conspired to make him a star. Archaeologists argued – in private, in public, in the press – about this remarkable man. Why was he up in the mountains, all alone? For a shepherd, he was lacking in traces of sheep; for a man with a fancy copper axe, he was a long way from the high-status burial places. What could have led him to die out there in the cold?

The cold had mummified his body, preserving skin, hair and even fingernails. At least 61 tattoos marked his shrunken and discoloured – but still present – skin. Researchers had everything they could possibly want to unravel the circumstances of Ötzi's death. However, it still took 10 years and one chance find to discover that the mystery of Ötzi's death was an unexpectedly violent one. Under the all-seeing eye of the X-ray, researchers at Italy's South Tyrol Museum of Archaeology, where Ötzi is housed, noticed something lodged just next to his left scapula. The foreign object turned out to be a flint arrowhead, rock-solid proof that Ötzi did not come to his end under his own terms. Shot in the back, he would have bled to death relatively quickly: the arrowhead was in position to have severed an artery. The evidence for murder mounted quickly: his hands had classic defensive wounds (cuts from fending off a knife), and a blow to the head had caused cranial trauma including a skull fracture. We have long considered Ötzi's end a sad one – on the run, suffering from wounds that would kill him in the cold and the ice. Now, even that is being challenged: recent work has suggested that he only ended up in the ice after being dislodged out of a proper grave by the forces of nature. His particular cold case is a startling example of how little we know about life in the past, and how long it takes us to build up a biography of bone, even when we have all the clues.

Interpersonal violence affects humans around the globe, regardless of age, sex or station. However, there are patterns to the violence that vary with time, culture and region. There seem to be patterns of non-lethal violence in the hunting and gathering groups of North America prior to the development of agriculture. The evidence of healed depressed cranial fractures from the Chumash and from other groups suggests that murder was not always the intent of one-on-one violence, but that interpersonal violence could vary considerably even when the end result – lumps on heads – was the same. It was adult males in the Chumash group who were most likely to have depressed cranial fractures, particularly those who inhabited the California Channel Islands, one of the more restricted-leg-room parts of the Chumash territory. By contrast, a group in Michigan had mostly adult females affected, which bends towards a theory of domestic violence against women. On Easter Island, another constrained environment, it's the adult males again, and among prehistoric Australians, women.

It's difficult to parse the evidence of interpersonal violence in the skeletal record. Lawrence Keeley, who wrote the influential book *War Before Civilization* (which we will come to in a few chapters), worked on the excavation of a Californian shell mound site in the San Francisco Bay; 30 years later, I had the same experience, albeit slightly further south in Malibu. In 2009 Robert ('Bob' in Keeley's preface) Jurmain published an update of the palaeopathology of the shell mound Keeley mentions in his book. The 'high' levels of interpersonal aggression in the San Francisco site resulted in a little over 3 per cent of the population showing skeletal evidence of head trauma. Phil Walker studied the frequency of cranial depressions in the skulls from further south in the Chumash area, including the Malibu region, and reported a rate of nearly 20 per cent; even higher rates were reported for Baja California, further south again. This level of violence doesn't sit well with the Chumash's view of themselves; it also seems to contradict a narrative of prehistory as a time of peace and plenty. But the bodies found along the coast of California have bashed heads and broken arms, and some are even studded with arrowheads; this is evidence that cannot be ignored.

Archaeologist John Robb of the University of Cambridge has looked at the rates of cranial fractures reported for the Neolithic period and the succeeding metal ages in Italy. He argues that while there are more cases of cranial trauma found in the Bronze Age, leading archaeologists to suspect underlying social causes of violence relating to the growth of complex political structures and urban life, the numbers of hits-on-heads is actually misleading. He argues that if you take the number of hits versus the number of heads, it's actually the earlier Neolithic stage that shows a 'spike' in violence. While it's hard to argue that the networks of villages Robb is describing in the Italian Bronze Age are truly 'urban', they exist in a landscape that is increasingly connected, and some of those connections do reach all the way to the populous, stratified, craft-specialising cities of the world. Cities by necessity must have highly integrated political systems to manage the long-distance trade and exchange networks of all the things a city needs – people, food and bling – and then manage the allocation of all those resources. Their innovative social structure bleeds out into the hinterlands in various ways. Robb sees the integrated social networks of dense settlements, usually based around hierarchies of males, as actually being *preventative* of violence. He points to anthropological evidence that for groups like the !Kung San (discussed in Chapter 2) or the Inuit, hunter-gatherer groups in very extreme environments, the homicide rate is actually very high; this is the same argument made by Lawrence Keeley, which we will discuss further in Chapter 8. In more complexly organised groups like the Turkana pastoralists of East Africa, homicide rates are lower than average, but a full three-quarters of adults surveyed in the late 2000s reported injuries due to violence, a majority of these being to the head. Ethnography paints a vibrant picture of both fatal and non-fatal violence, with more deaths at the extreme ends of population density.

In China a survey of fracture patterns showed that the agro-pastoralists of the Mongolian frontier under the thumb of the Chinese state during the Late Bronze age actually had *fewer* fractures than nearby free-range nomadic pastoralists; to be fair, though, they also had iron helmets. A sample from the site of Lamadong in the early days of Imperial China, a proper nation-state if ever there was one, has fewer still. There is a school of thought that, because non-agricultural groups in the modern world are invariably in some sort of quasi-subjugation to the political forces of the world of nation-states, any violence observed in hunter-gathers is a pathology introduced by contact with Western/state/military/capitalist (delete as appropriate) forces. The skeletal record would beg to differ.

Robb sees the number of males with cranial trauma at the Italian site of Pontecagnano during the Bronze Age as signalling a lowering of interpersonal violence; city life offered a more refined way to settle arguments than blunt force. The Neolithic he finds more equal, with both men and women experiencing similar rates of cranial trauma. The rise of the tools of violence as an important cultural symbol, reflecting particular social identities and status, might actually reflect the real use of weapons in a society as symbols: not so much to kill, but to suggest the status of a killer. While his example comes from one particular site in a particular period in a particular region, it's interesting to try to extrapolate these patterns beyond Italy. Up in the Alps, the murder of Ötzi fits into a culture and a time of Neolithic competition in the Alps that these theories suggest should have been less lethal. However, lethal and non-lethal patterns of violence can be found in many places around the world where a combination of life tied to the land and social stratification intensify the potential for conflict. Ötzi's ravaged remains tell us at least one aspect of the story of the Neolithic that we would see in the Americas, and probably Asia and Africa too, if only we had better preservation.

But the Neolithic is the time *before* cities, and what we really want to know is if the brand-new urban systems that follow the development of settled, agricultural life show the same calming effect as the less-urban networks of towns and villages that Robb describes for Italy. I should reiterate the point about metal not being the harbinger of cities: much of Europe adopts metal without getting very populous and urban about it for thousands of years. The Italian Bronze Age is not quite the same as the Mesopotamian one, and just because I'm cavalierly referring to various metal ages to keep my dates straight doesn't mean that the hoary old progressive trajectory of stone–bronze–iron–steel is applicable in even the majority of past cultures. China, for instance, managed perfectly good cities with jade at the top of their must-have list until truly global markets encroached in the post-medieval period, and the Maya never had a Bronze Age at all but certainly had cities. One of the great tragedies of archaeology is that the very first cities were largely the ones in places archaeologists' governments had just colonised, so many of our terms – especially those for time periods – are unfairly coloured by the well-known archaeology of the ancient Near East.

While modern-day excavators may insist on a variety of ridiculous and anachronistic practices and tools,* the science has moved on considerably since the days of pith helmets and cocktail hours.† Where once the aim of excavation was to blitz down to the bottom of whatever pile of archaeology you had,‡ yanking out the interesting bits§ as you went along and hawking your story and the goods at the end, now we have Harris matrices, finds registers and fine-tuned public engagement strategies. We are, however, still lumbered with some of the highly specific, non-universal terms for different time periods that rose out of those mud-brick mounds. Hopefully the reader will have the fortitude to wade through the morass of dates, times, periods and technologies and accept my broad divisions of the world into the non-urban Neolithic of villages, the urbanism of early cities and (when

* Gin, short-handled shovels.

† We've given up the pith helmets.

‡ Apparently, laying light rail *through* a site to get the finds out was considered acceptable.

§ Read: gold.

we get there) the global network of a truly urban world of connected cities, empires and states.

As something of an illustration of the widely variable experience of the interpersonal violence we have been discussing here in the past, the mighty site of Harappa in the Punjab in modern-day Pakistan existed at exactly the same time as old Ötzi, but in a radically different milieu. More of a network of villages in Ötzi's day, by its peak in the mid-second millennium Harappa was a proper urban city containing an estimated 20,000 people. The culture that occupied the Indus Valley in the third and second millennia BC, stretching across much of the subcontinent and trading with the city-states and nascent empires of the Middle East, was once idealised as a peaceful, stately exemplar of humankind's path towards civilisation. Because the site had long been associated with the semi-mythic theories of an Aryan Invasion, it seems that perhaps researchers had gone slightly too far in debunking that particular story of violent conquest and might have pushed interpretations of any kind of violence out of the picture. Harappa was the rainbow-tailed unicorn of the human past: an unstratified, largely equal society, whose road to complex urban society had been paved with non-violent interactions.

One person's unicorn, however, is another's devastating blunt force trauma to the skull. Bioarchaeological research has shown that not everyone living in the huge early urban site of Harappa had an easy time of it. A handful of such individuals (a few more of them female than male) were identified from the mixed bag of human remains excavated over a near century's worth of archaeological digs. Under the unhelpful glaze of archaic preservative that early excavators dumped on the human remains, there were clear signs of painful and sometimes fatal cranial injuries. The number of violent injuries exceeds both the preceding, non-urban phase of the Harappa settlement as well as the record of violence from contemporary villages and isolated skeletal finds from the same region. Broken down into different periods of the city's history, however, the evidence of interpersonal violence tells a slightly different story. The markers of violence from the bones buried at the height of the urban era pale in significance compared to the evidence of the post-urban period, when social collapse and instability seem to have upped the stakes for everyone. In this case at least, urban living lowered your chances of being hit upside the head.

What seems clear from the archaeological record is that with the end of the Bronze Age and the beginning of the Iron, the symbols of war start to be backed by the actual physical evidence of war. So while we might think of the first stages of urbanisation as actually having a dampening effect on the sort of face-smacking methods of argument resolution used in prehistoric California, it seems that after a certain point, just walking softly and carrying a big spatula is no longer enough to prevent violence. When we reach the point of proper-sized cities, with their endless need for goods and status and their endless supply of weapon-ready hands, we invent a whole new set of reasons to kill.

CHAPTER SEVEN

Under My Thumb

In 1969 sociologist Johan Galtung published a paper called 'Violence, Peace, and Peace Research'. In it he laid out what he understood of violence, and the definition he gives is very different to the one we have been considering in the previous chapter in terms of blows landed and enemies dispatched. He extends the idea of violence to the circumscription of what people can do, achieve, or be.* In his example, violence is implicit when bad things happen and they did not have to; death from tuberculosis before the advent of modern medical treatment would not be violence, but death in the modern world, where treatment exists, is. Someone, somewhere, is being prevented from being cured, and that is an act of violence. It's a sobering thought, but begs the question of who implements this kind of violence, and how. In Galtung's formulation, there is a line drawn between a personal act of violence and one in which there is no clear actor; the latter is what we mean when we talk about *structural* violence. This chapter discusses the violence in between the personal and the professional; direct physical constraints on physical lives. This is the violence of society enacted on its members, of human processes on human beings. Much of what Galtung considers structural violence can be found in other parts of this book, particularly the sections on inequality, but here we will restrict ourselves to physical processes enacted on entire groups of people that we can identify in the bioarchaeological record – systematic physical violence directed towards specific groups in society: women, children, social inferiors, criminals, prisoners of war and outsiders in general.

---

* He actually says 'prevention of realisation', but I'm trying to avoid the less sociologically inclined reader from actively contesting this definition of violence by chucking the book across the room.

If we are to talk about violence incurred by vulnerable groups in society, one of the most depressing places to start is with a group that is, by definition, dependent on carers for survival: children. Child abuse, as defined by UNICEF today, is a very tricky concept to apply retrospectively, including as it does: 'all forms of physical or mental violence, injury and abuse, neglect or negligent treatment, maltreatment or exploitation, including sexual abuse.' As with domestic abuse, specifically violence against women, societal attitudes about what constitutes 'abuse' have changed markedly.* In several countries it's illegal to physically discipline a child, while in rather more it's an expected, accepted part of childhood.† Archaeology can be frustratingly obtuse on the subject of the patterns of discipline considered socially acceptable in the past, but bioarchaeology, on the other hand, can give us clear evidence of children who were sufficiently mistreated that they bear the scars on their very bones. Scarred children are not a long-term survival strategy in any evolutionary context, but given what we know about modern-day rates of child abuse,‡ it's surprising that we don't see much evidence of child abuse in the skeletons of the children of the past.

Of course, there are many reasons for this, and not all of them evoke a better world. In the previous chapter on violence, we have discussed the bony response to trauma in the case of fracture. We also covered the special case of greenstick fracture in childhood, and how the less-mineralised bone of the still-growing child responds to force slightly differently than adult bone. The plasticity of juvenile bone, and its capacity for remodelling, may be one of the main factors in masking our awareness of children's broken bones in the past; healed breaks that have been survived for some time might be virtually invisible to bioarchaeological

---

* Example: If you are of an age to have loved the film *Sixteen Candles*, you may have missed the horrifyingly date-rapey subtext of basically the whole plot. Times have changed. Mostly.

† Having never misbehaved as a child, I would not know.

‡ Do *not* look this up if you want to sleep at night.

investigation. The clinical definition of Battered Child Syndrome requires evidence of *chronic* mistreatment: a pattern of healed, healing and new injuries. In modern instances this can be identified in soft tissue damage as well as in the skeleton, but for cases in the past we have only the bones to guide us. The bones of adults may not transmit a true signal of the trauma experienced in childhood – bones and bruises, at least, might heal.

In the remains of children, however, a combination of evidence of bruising and fracture can be used to build a diagnosis. Violent trauma, direct blows or the grabbing and manhandling of limbs can injure the sausage-casing-like periosteum around bone; this leads to 'bone bruising', in the words of our constant companion in the world of bioarchaeological violence, Phil Walker. The damage to periosteum stimulates bone activity to the extent that an actively healing area of injury can be identified by the characteristic new bone formation sequence on the hard outer surface of the bone: scattered tiny dots that indicate microscopic pitting and thin straggly lines of new bone forming on top of old. This same sequence occurs with 'anything that tears, breaks, stretches or even touches' the sensitive periosteum; the whole inflammation routine, however, can be triggered by any number of conditions, including infectious disease and parasites. When observed in combination with fractures, however (and particularly the typical fractures of the ribs or ends of growing long bones that result from shaking), it may be possible to identify cases of child abuse in the past.

The large cemetery at the site of Kellis 2 in Dakhleh Oasis in the middle of Egypt's Western Desert gives us an example of what evidence of child abuse might look like archaeologically. Dating to the Roman period, sometime in the first 500 years AD, a two- to three-year-old child died with considerable trauma. There were broken ribs, 'bone bruises' to the arms, broken arms, a broken clavicle and even fractures to the vertebrae and pelvis. These injuries were at different stages of healing when the child died, indicating a pattern of trauma indicative of child abuse. Mary Lewis of the University of Reading has compiled the known instances of violence against children in the archaeological record – a mammoth task, but surmountable because the numbers involved are actually very small. Part of this, as Lewis argues, is the above issue that the bones of children do not respond to trauma in quite the same way that adult bones do, so we are looking at a very different reflection of violence in the past. That said, only a very few of her already limited number of cases can be interpreted as evidence of child abuse. In Europe, a total of three children have typical fractures of the limbs and metaphyseal growth plates mentioned above, one from Roman France and the others from England in the Roman and medieval period. Is this proof that the past was kinder, and children more loved? Or just a reminder that children's skeletons are plastic, growing things and that the skeletal process of remodelling can remove traces of violence from archaeologists' eyes?

These are isolated cases, which Phil Walker has argued is evidence that child abuse was in fact very rare in the past, despite the reputation of 'old-fashioned' methods of child rearing; even then it's the physical discipline of school-age and above children that most commonly comes to mind.* He argues that it's only our modern urban settings, divorced from normal social networks of a more integrated family life, which encourage the pathologies of child abuse. The skeletal evidence of violence that we see in modern forensic cases of child abuse should be visible in the past, if rates were as high as they are today. In Walker's conceptualisation, urbanisation means anonymity, and isolation can be maintained without family or other parties becoming aware of the abuse and stepping in. I would agree with him that there are many factors that might be masking abuse in the past, predominantly to do with the poor survival of child remains, the potential for evidence of childhood abuse survived to be remodelled and absent at time of death, and even, I would add, where the

* Belts, first and foremost. Never give your grandparents reason to discuss how they were raised.

victims of child abuse might be buried and whether or not archaeologists would find them there. I'm not so convinced, however, that the isolation he suggests was a real possibility for most urban dwellers. If isolation was a factor, then the high mobility and nucleation of family units of the post-war period of the twentieth century might seem to be a more immediate cause. But we know that child abuse did exist before extended families died out. There are certainly records from the medical establishment, unheeded at the time, of a plague of child abuse from at least a century before the identification of Battered Child Syndrome in the 1960s. While it seems that the horror stories related by our own parents (and theirs) of the cruelty our (your) generation has so narrowly escaped may have been slightly over-egged,* there is potentially an argument to be made from the paucity of evidence of child abuse in the past, and its reappearance in modern times in situations of social disintegration, that violence against children is an extreme experience, and not the norm in town or out.

The second type of structural violence against vulnerable individuals is sadly very much a normalised pattern of abuse that seemingly stretches back as far as history takes us. The physical abuse of women seems to have occurred throughout the past and in cultures from all over the globe. As touched on in Chapter 6, there are specific fractures and wounds that are associated with self-defence. When associated with evidence of the head trauma that the defence was guarding against, a common bioarchaeological interpretation is that the victim was defending against attack. One could argue that the fractures and other evidence of trauma on the skeletons of women might be just as easily attributable to accidents, or even a more straightforward episode of *mano a*

* For instance, some of the recorded punishments we know of from the writings of the Dutch urban elite over last 500 years include things like the cruel, yet patently familiar, sent-to-bed-without-supper ordeal.

mano violence, but one would be talking absolute bollocks.* There is an extraordinary amount of literature that covers the ethnographic, historic, anthropological, psychological and sociological evidence of violence against women, and while I haven't got a snowball's chance in hell of covering most of it here, I can helpfully summarise. Violence against women is endemic. It occurs in all types of societies, at all levels of complexity and at all times. While bioarchaeology may not be able to distinguish exactly those episodes of violence that are the result of structural features of society and those which are not, modern statistics can give us some guidance: according to the World Health Organization, one in three women will be victims of domestic physical or sexual violence in their lifetimes.

In the case of domestic abuse, Phil Walker's argument for the causes of child abuse – that modern urban society is responsible for the rise in cases – falls flat. We can take, for example, the !Kung San people, first encountered in Chapter 2. The first ethnographers of the San were amazed by their relatively low workload and seemingly peaceful attitudes. The observations by anthropologist Richard Borshay Lee in the 1960s featured the shocking revelation that hunting and gathering appeared to be much easier than it looked once you knew what you were doing. Further research decided that men worked more than women,† society was very egalitarian and violence was largely absent. It might be helpful, for reference, to know that these observations were made in the context of what we might call the 'yoghurt-crocheting' zenith of cultural anthropology – that they were made largely by men, and that they were made by members of a society

* My language here is why Millennials have invented the expression 'sorry/not sorry'.

† Borshay Lee judged that men did more work than women by counting up time spent hunting (men), fixing stuff (men) or gathering (women). Literally nothing else is considered. Guess that water just up and gets itself, and those kids just magically grow.

that still had[*] systematic legal discrimination against women. Otherwise it's very difficult to square these practical observations of broad universal themes of egalitarian equality with the personal narratives of women of the !Kung San. In the early 1980s, anthropologist Marjorie Shostak published a book focusing on the accounts of Nisa, an actual woman of the !Kung San. Nisa confirmed that life among the 'egalitarian' hunter-gatherer tradition included considerable amounts of domestic violence against women, including memories of her father kicking a pregnant wife, and her own daughter's death due to a violent husband. These personal tragedies occur outside of a world of urbanism and cities, but they do fit with one of the suggestions that Phil Walker made about child abuse: that it happens in times of intense social stress. The integration of the last hunting and gathering tribes of the Kalahari into an agrarian state culture has not been a bloodless transition, and it might be that Nisa's story is part of a larger violence affecting the entire social structure of the !Kung.

When we look at archaeological evidence for periods when domestic abuse (or at least violence against women) seems to increase, there are some tempting patterns. In the Late Woodland period, which sees the slow build-up of maize-growing and sedentary life in parts of the state of Michigan, the site of Rivière aux Vase contained considerable evidence of cranial trauma. Even greater numbers of heads were smacked than in regional contemporaries, and again, the blows fell on more female heads than male: 13 per cent of the women's skulls examined showed evidence of healed depressed fractures. In Neolithic China, nearly 40 per cent of females in one period at Jinggouzi had broken noses; it is not known how representative a sample this group was, but it's still an astonishingly high rate. In the Pueblo cultures of the American Southwest, a full 60 per cent of the women of the relatively comfortable Neolithic communities of the La Plata region have evidence of cranial trauma, versus 23 per cent of the men. In many of these situations there is some

[*] Has.

suggestion that environmental change (*e.g.* drought in the case of the Ancestral Puebloan culture of the American Southwest) was responsible for causing society-wide stress, but not all. Despite an overall trend for male skeletons to bear more evidence of violence, accidental or intentional, there are significantly more cranial fractures among the women of an agro-pastoralist Bronze Age population who lived under state control along China's northern Mongolian frontier at Jinggouzi than among the more free-ranging pastoralist groups beyond.

Who else can a society turn its ire on, once we're done with the women and children? Ever more specific groups that somehow contravene or subvert social order, people who mess with the natural order of the universe and cause pain and suffering for the rest of society. And it turns out that two of the groups most often accused of interfering with the natural order of things happen to be the marginalised majority: women and children. There is no scope in this book to go item by item through the list of oppressed minorities. We can, however, try to bring out the bioarchaeological evidence for what has happened in the past, but with the understanding that it's very difficult to pin down a diagnosis of structural violence without the evidence of population-level trends. So while we can examine rates of violence against women and potentially recognise signs of child abuse in the past based on bones alone, we will need further evidence that other groups were intentionally targeted for maltreatment or even death by the societies in which they lived.

This is not to suggest that groups defined by different mental or physical health statuses and abilities, gender performance and sexual preference, or any other trait, were not targets of structural violence in the past.* It is merely to say that we need to have a way of identifying signs of violence against them from an archaeological perspective. One of the primary sources of evidence for differences in treatment in life is difference in treatment at death; this is encapsulated in

* Because they sure are in the present.

the term 'deviant burial'. Where there are cultural norms for burial – neatly laid out versus messily heaped, not-squashed-under-a-rock versus squashed-under-a-rock – we can identify deviation from those norms. The dead mean most to the living, and the treatment and curation of their remains gives us an idea of how the society that did the burying felt about the individual that did the dying.

We can further separate deviant burials into types by how they deviate from the norm. In the archaeological evidence, the clearest indication of separation from the community in life is separation in death. Bodies buried in isolation (or simply not buried at all) or remains found mixed into trash heaps and lining ditches may signal that their owners were not necessarily valued members of the community. This treatment isn't necessarily reserved for social deviants. About a decade ago, I ended up staying in the tiny village of Mitata in the Ionian Islands by the grace* of the Kythera Island Project survey, the slightly bigger precursor on a slightly bigger island of the Antikythera Survey Project mentioned in Chapter 3. Despite being engaged in a very different project,† I eventually got called down to the ferry harbour at Diakofti to come check out some bones found by the field school run by the indefatigable, much-loved local archaeological authority Aris Tsaravopoulos. I should say that Diakofti is a special place by any account: a minute village of four streets set firmly into a steep, featureless hillside, it only got running water in the 1980s. The place exists more or less solely because at some point the main port was relocated to the crystal-clear waters of its relatively large harbour – not clear enough, however, to prevent a drunk Russian sea captain from very scenically beaching a tanker just on the other side of the causeway connecting the harbour to the island proper. The marooned vessel formed a fitting backdrop as we bounced the rental car over the causeway and onto the undeveloped land facing out

* Read: spare bed.

† Writing the methods chapter of my PhD. It seemed better to do this on a Greek Island than a dank London flat, for some reason.

to sea.* There, the mildly grubby faces of the field school students looked up expectantly from their latest find – an isolated grave, alone on the shoreline, facing out to sea.

I'd come to look at the body that had been exposed, and to settle whatever arguments remained among the archaeologists. Good money – *i.e.* unfounded gossip – suggested that the interment had been of an outcast girl, a teenager, who had died of syphilis sometime in the 1800s. This was a problematic interpretation as the skeleton was robustly male, with big limbs and pronounced chin and brow line. There were no signs of syphilis (discussed further in Chapter 11), but there was a potentially broken leg and quite a bit of reactive bone growth around the holes of the ears. I disabused the archaeologists of their favourite working theory and went on my way. Some weeks later, as yet more gossip spread around the town, I learned the sad truth. The locals did indeed know about the burial on the shore, and had for a very long time. The man in the grave by the sea had been a sponge diver from Corfu – hence the bony changes around his ears, caused by inflammation relating to the constant change of pressure† – and when he died, no one in the village had had the money or will to send him home. So his burial occupied a funny halfway house between uncaring disposal and proper burial in the village churchyard.

In a similar vein, many readers might be familiar with the superstition about burying suicides at crossroads: it was in fact British law until 1823. Strict rules guarded the sanctity of the burial ground, and even after the crossroads thing was given up, suicides were supposed to join the paupers and the unbaptised. Of course, we know that these rules were regularly flouted. At St Bride's Church, the fabulous Fleet

---

* Awkwardly, we did this in full view of Philipos, the person who had rented us the car. The one cafe in Diakofti sits at the end of the causeway and doubles as his second office. This pattern of behaviour may explain why so much archaeological field time is taken up with car trouble.

† Getting that diagnosis correct was a serious ego-puffing moment.

Street institution where the pews are all sponsored by tabloids (it is traditionally the journalists' church), a small collection of remains were recovered in neat coffins, complete with metal plates giving the occupants' names and occasionally ages, after the destruction of the Blitz in the Second World War. These remains were carefully stored in a locked crypt for around 200 years before the bombs reopened their tomb, and after initial excavation have remained housed under the care of the church in which they were originally buried. Among these remains is the sad case of a young man, a young husband with a respectable career, who committed suicide by gunshot in 1821, before the crossroads law was even abolished. Newspapers reported the circumstances of his death, so it was hardly a secret; however, despite his mortal sin, he was still interred alongside the rest of his parish, with the record of his burial noting only that he was 'suddenly found dead'.

Through a combination of bones and burials, we have access to clues about structural violence against people perceived to contravene social, cultural or spiritual norms from the way their bodies are treated in death. First there are the bodies of the morally, spiritually or otherwise dangerous: these are bodies that must be buried with special precautions. Then there are the bodies that simply are not worth the effort: remains of people whom the ones doing the burying don't really consider people. There are also the bodies of those who require punishment even in death: the Anglo-Saxon tradition of burying executed criminals in embarrassing positions* is one of the best-known examples of this. Finally, we have what may be the most flagrant abuse of bodies by society: taking either life, liberty, the body or its constituent parts and using them for your own ends. Into these categories, we can fit among the dangerous: suicides, witches and criminals; among the uncared for: enemies, the poor and outsiders; and among

* As exemplified by my first question to an archaeologist presenting some of his work on this: 'The hand was found *where*?'

the ones whose very bodies are used against them: a motley crew comprising slaves, prisoners of war and human sacrifices.

First, let's talk about witches. Witchcraft is remarkably common, and not just at your local folk music festival. Witches are one of the primary causes of bad luck in many parts of the world: sorcery, black magic and other acts of spiritual malfeasance can explain everything from crop failure to disease. It doesn't particularly matter whether witches are real or not; in societies where a belief in the possibility of black magic or harmful spiritual activity exists, there are individuals who are accused of directing it. The UN Commission on Human Rights estimates that more than a thousand people are killed for witchcraft every year in Tanzania alone; while 'witches' are regularly killed in Asia, Oceania and Africa. In my own London neighbourhood, I receive three or four business cards a year, slipped through the mail slot in the front door, advertising the services of spiritual healers who can detect black magic done against me, not to mention advise me about marrying well, money and other spiritual matters.* These cards are part of a wider narrative that includes the resurgent practice of exorcisms and church-sanctioned violence, and culminates in the modern practice of 'necklacing' suspected witches – dousing old tires in gasoline, sticking the witch in the middle and lighting a match.†

By and large, in culture after culture, witches were elderly women. It is they who bear the brunt of witchcraft accusations in the modern and ancient worlds, whether as an escalation of disagreements between community members or because they are a handy scapegoat. The question remains, however: how do you find witches once they've already gone to ground? Only a handful of archaeological cases exist that have been

* For instance, the controversial preacher Helen Ubaiko says she can save you from witchcraft, ancestral or mermaid attack. Critics say she encourages abuse of children who are deemed to be possessed. I personally would like to see the mermaid up front before handing over any of my money.

† Witnessing the aftermath of such an occurrence gave one acquaintance a lifelong fear of the rural part of the African country he was working in.

identified as 'witch' burials: a teenager under a rock in Lancashire; several 'prone' burials (where the deceased is placed facedown) from across Europe; and possibly a woman who died of lead poisoning on a slave plantation in Barbados and who was buried in a special 'mound' grave above the normal level of the earth.

There is good reason to suspect that in many cases, even people killed for witchcraft still received a fairly normal burial: several of the people hanged in Salem, Massachusetts during the sixteenth-century witch trials were later exhumed and buried elsewhere by their families. When archaeologists went looking for bodies under the infamous Gallows Hill in Salem, they found no evidence of any of the witches that had been supposedly buried there. In the absence of bodies under a gallows, the only way that their deaths could be identified as abnormal would be through the evidence of their bones. Judicial killing by 'long drop' hanging can cause a 'Hangman's fracture' a characteristic fracture to the second vertebrae from the skull – the hyperextension of the neck by the suspension of the weight of the entire body. However, forensic research on the exhumed bodies of criminals executed by hanging in the nineteenth and twentieth centuries found the fracture in only 3 out of 34 cases, so even then we would have a hard time identifying them.

The rules that apply to witches apply to criminals too – deviant behaviour is rewarded with deviant death and burial. Aside from the hangman's fracture discussed above, there are a number of methods of judicial killing that we can identify in the past. Judicial killing, or the justified imposition of death on an individual according to established and agreed rules of society, may occur anywhere, and at any level of complexity – at some level, even witches were 'rightfully' killed in the eyes of society, whether or not condoned by an arbitrary representative of a state or hierarchical power. Judicial killing can take many forms, most of which seem to have been perfected at one point or another by the Inquisition, though the Romans should take their fair share of the credit. The Roman punishment for parricide – killing your parents – seems to have been designed both with the goal of protecting parents from mercenary heirs as well as with a degree of whimsy. The convicted parricide sentenced to *poena cullei*[*] – they were condemned to die in a sack, specifically an extra-large sack the size of one oxhide sewn together. The sack would be filled with the condemned, a dog, a rooster, a snake and a monkey, and chucked into a body of water. What symbolism was intended is lost in the mists of time, but it certainly captures the attention. No such commingled remains have ever been encountered in the beds of Roman rivers or lakes, so archaeology awaits further information on how much practical use this type of execution actually got.

Crucifixion, interestingly, is another type of judicial killing that you would think we would know more about. It was certainly popular enough. One of the problems with identifying crucifixion in the archaeological record perhaps comes down to the usual problem of not finding the bodies that are buried away from the rest of society, but may also be explained by the fact that we are not terribly sure how crucifixion was actually done. Any good Christian[†] will have somewhere a mental image of *one type* of crucifixion, but we know from the writings of Seneca that there were multitudes of ways to die on the cross. The cross, for one thing, didn't have to be of the sort represented by the Christian crucifix; it could be more of an 'x' shape; the victim could be attached to it by nails or rope, which could be around or through hands, arms, one foot or both, or even the groin:

> I see before me crosses not all alike, but differently made by different peoples: some hang a man head downwards, some force a stick upwards through his groin, some stretch out his arms on a forked gibbet.
>
> *Seneca, 'On Consolation to Marcia', Chapter XX*

---

[*] Literally, 'penalty of the sack'.

[†] And one terrifying pre-school teacher with a predilection for dioramas. How was a non-Christian kid supposed to know what to make with the two Popsicle sticks?

The only bioarchaeological evidence of the practice in existence is a solitary heel bone, uncovered in Israel and controversially identified as a belonging to a victim of crucifixion because of the long nail perforating the middle of the bone. Potential supporting evidence comes from the fragmentary tibias in the same skeleton, which could have been broken according to the common practice of smashing the legs of the condemned, designed to either increase suffering or hasten death.* Bioarchaeologist Kristina Killgrove has pointed out in her regular column for *Forbes* magazine that there could be a variety of reasons why we don't have more evidence for crucifixion in the archaeological record, ranging from the reuse of crucifixion nails as spiritual objects to the likely discard of criminals' bodies in less salubrious locations. It may be that the thousands of victims of crucifixion are simply lying undiscovered somewhere, far away from the cemeteries and graveyards of law-abiding Romans.

As for judicial killing practices that are more easily identified from the archaeological record, both beheading and 'drawing and quartering' are practices well known from TV and movies.† Butchery marks at the appropriate locations are the key bioarchaeological evidence in these cases. Just as your steak knife leaves little indentations on the roundel of bone at the centre of your beef, hacking off a head leaves incontrovertible cuts, gouges and chop marks. Mary Lewis has identified the only known evidence of a skeleton of a man who appears to have been hanged, drawn and quartered, though in this case the powers that be seemed to have decided to throw in emasculation and beheading free of charge. In the time-honoured fashion of medieval tragedy, the story starts with royalty. In the 1300s, Edward II of England took the son of an extremely wealthy and powerful family, the Despensers, as a lover. Hugh Despenser turned

---

* Or both.

† Props to Ned Stark for the opportunity to demonstrate the verb form of 'beheading' in both active and passive case.

the king's head but more definitely put Queen Isabella's nose out of joint. After a period of deteriorating personal and political relationships, Isabella went to France, found herself a lover and an army, and came storming back across the channel in 1326. She caught up with Despenser in Hereford, and executed him according to the new fashion that Edward II's predecessor, Edward I, had come up with specifically to deal with Welsh separatists. Adding to the traditional practice of being drawn (tied to a horse, dragged through town) and then hanged, with the bodily remains quartered or split into four parts and dispersed, the condemned would be disembowelled and beheaded after the hanging. The account compiled by Lewis of Despenser's last hours on earth suggests that the death was not an easy one. First, Despenser was humiliated, the sigils of his house symbolically reversed, and he was given a crown of nettles to wear. He was then dragged by four horses through the streets of Hereford, stood up on a ladder some 15 metres (50 feet) in the air and 'hanged' on a special gallows in front of the walls of his own castle. Choked but still conscious, he had his genitals cut off and burned in front of him, before he received a knife to the chest and was disembowelled. His entrails were burned, his head cut off, the rest of him chopped into the appropriate number of pieces, and Isabella's justice was done. It's a remarkable piece of bioarchaeological detective work, consolidated by the clear evidence of the bones: the upper vertebrae are sliced through the middle, there are chop marks on the clavicle and arms, and even a stab wound that has gone through to the inside of the chest and left its mark on one of the vertebrae.

The execution of Hugh Despenser was specific enough that the trauma it inflicted on his body could be traced back through his bones. Beheading, on the other hand, is trickier. Coming back to the concept of deviant burial, we require a combination of bioarchaeological assessment of whether the individual was alive or dead by the time their head came off, as well as knowledge of the cultural context of their death, to understand whether an isolated head is evidence of structural violence.* Many cultures, starting well back into the Neolithic, maintained skulls as some sort of ritual practice, as evidenced by the 8,000- to 9,000-year-old plaster-coated skull with shell-inlay eyes found at Jericho and now in the British Museum. Decapitation might be part of burial rituals, or it might be a judicial killing; without context it can be difficult to distinguish between symbolism and sentence. Beheadings certainly have been used as means of judicial killing,† but they have also been used as means of protecting against spiritual harm from 'dangerous' burials; cutting the head off suspected vampires falls into this category. Unfortunately, the bioarchaeological evidence of vampire burials – decapitation, piercing wounds, unusual body layouts – rather exactly matches what might be expected from the burials of witches, executed criminals and even suicides.

The case of Hugh Despenser is illustrative of the emotive power of the human body, both in the manner of death and in the disposition of it afterwards. In the not-so-distant past, those on the losing side of conflicts not only suffered the inconvenience of dying, but the further indignity of their physical bodies being captured by the enemy. While we in the modern age may attribute our spiritual being to a nebulous ephemeral realm and have no compunction about burning our corporeal remains to ash, this is not a universal concept. For a great deal of known religious thought, past and present, the body is crucial to spiritual life. It is not hard to imagine that people in the past would have been horrified (as all but the most

* For instance, museums all over the world are full, absolutely full, of skulls without bodies. This has a lot less to do with decapitation, however, and a lot more to do with the limited interest anthropologists of an earlier era had in anything that wasn't constructing skull morphologies and racial typologies. Not that we'll know if any of those isolated skulls were decapitated: the necks are missing.

† Ask the French.

sanguine of Gnostics would) by the thought of their heads on spikes as trophies of war.*

If we want to think about the abuse of bodies, in life as well as death, there can be no better example than empires. Complex urban political systems, displaying their power to an audience both at home and abroad, seem to provoke a specific type of violence that emerges alongside those very first cities. It precedes and then pervades the empires of the world: in both of the Americas, in Europe, in the Ancient Near East, in Asia, in Africa. It can be a form of structural violence, the violence of society against its members, or it can be a form of aggression turned outwards, but it leaves bodies in its wake just the same. If any kind of death can be attributed to the dawn of cities above all others, it's the very particular case of people killed as chattel – as gifts, as pets, or as subjects. The rise of cities is intricately tied to an entirely new sort of power – the power over individual lives. This is the most revolutionary aspect of urban living: a concentration of power that extends to the control of the physical bodies of the population. There are two principle expressions of a state's control of the physical beings of its subjects: control of labour, and control of life. The former might take the form of corveé labour, obligatory communal agricultural work or even forced factory or specialist labour. The latter is more accessible through the skeletal remains of the physical bodies so manipulated: not only judicial killing, but evidence of mutilation, and evidence of human sacrifice.

We know quite a bit about human sacrifice in the age of empires, particularly from the Americas. The Olmec, the Aztec, the Maya, the Moche, the Inca – these big prehistoric states all utilised human sacrifice to build and maintain their relationships with gods and men. The Moche, perhaps

---

* And I bet you even they would blink. Whether this was actually empirically tested by the Inquisition in the thirteenth-century Albigensian Crusade remains unknown, but it was certainly a scientific inquiry taken up with gusto by the time of the French Revolution.

Left: The reconstructed Neolithic village at Aşıklı Höyük. The only concessions to modern sensibilities are the gaps between buildings; 10,000 years ago Aşıklıs inhabitants would have used the roofs as streets.

Left: Also at Aşıklı Höyük. Buildings are made with locally sourced mud, using traditional vernacular architecture from villages around the site.

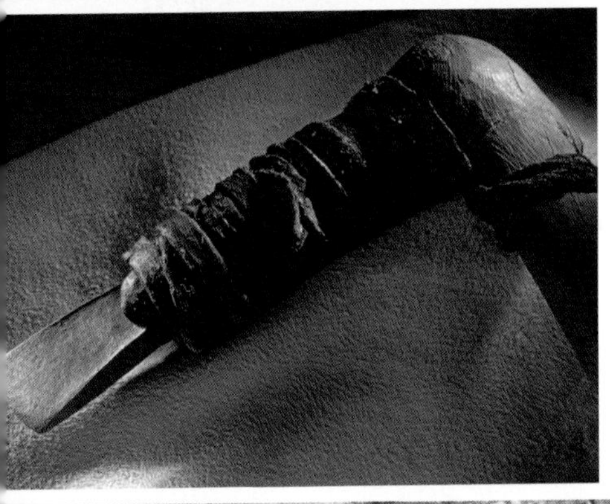

Left: Ötzi the Iceman's re-hafted axe. Archaeologists have argued this weapon is more important symbolically than in actual conflict.

Below: Ötzi thawing out of the Alps in 1991. The preservation of the body allowed researchers to reconstruct his grim last days.

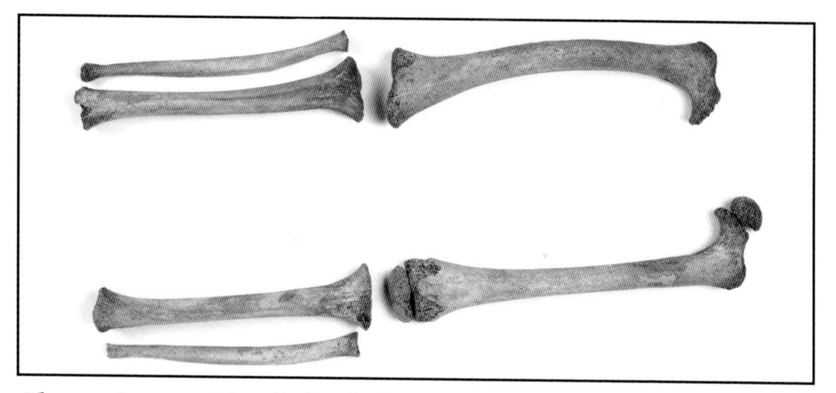

Above: Some of the skeletal changes associated with rickets. The lower limbs of a child (note the 'unfinished' appearance of the ends of the bones) show considerable bowing.

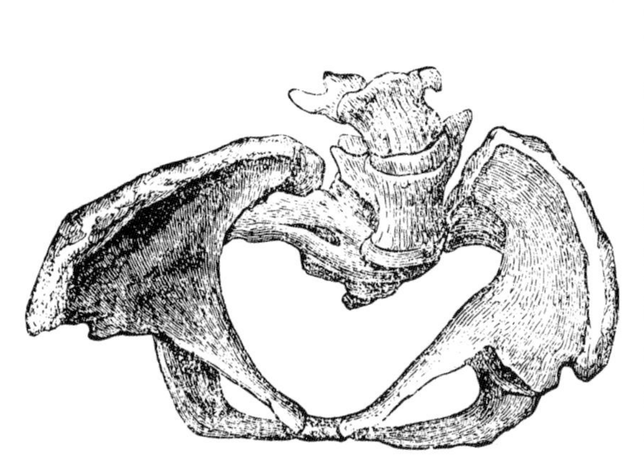

Left: A rachitic pelvis. The birth canal shape is affected by the change in angle of the bones, potentially proving fatal in childbirth.

Below: Excavations at Başur Höyük uncovered remarkable burial chambers like this one, where goods were piled up beside the tomb's occupants.

Above: The amount of bronze finds at Başur Höyük were astonishing. There were so many spearheads in one grave that they had to be tied up in bundles. Traces of this wrapping is visible on the spearhead shown inset.

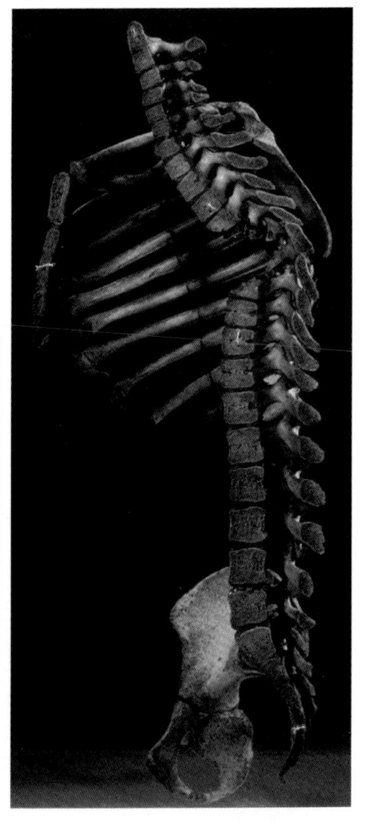

Right: The effects of infection with *Mycobacterium tuberculosis* can be seen in this severely angled spine. This is 'Pott's Disease', or angular kyphosis.

Below: An image from around 1200 AD shows a leper with disfigured facial skin.

Left: The world's first historical document, the Stele of the Vultures, commemorates a war between two cities. The god Ningirsu holds a net full of enemies – humans could be considered spoils of war.

Left: The protracted execution of Hugh Despenser in Hereford in 1326, on the orders of Queen Isabella of England. Despenser was drawn, strangled, stabbed, emasculated, disembowelled and beheaded before being quartered.

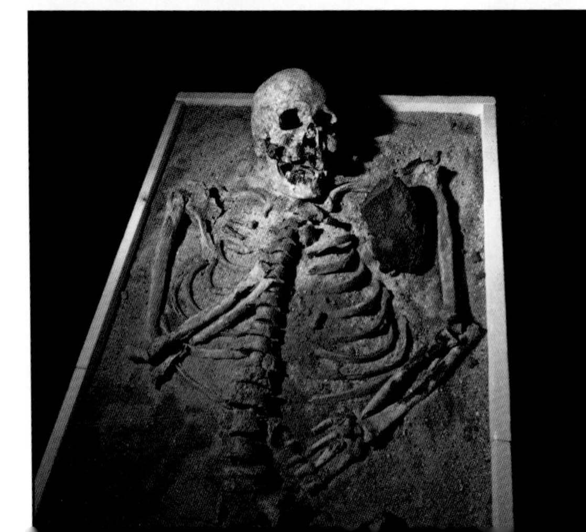

Left: A deviant burial from medieval Sozopol, Bulgaria. It is likely the deceased was 'staked' with iron, a practice commonly used to guard against vampires in Eastern Europe.

Above: Match factory workers on strike in London, 1888. The women workers of the Bryant and May Match Factory were subject to abominable working conditions, including exposure to the white phosphorus type of matches that caused 'phossy jaw'.

Below: The jaw of a man, probably a matchmaker, with considerable destruction due to tissue necrosis caused by 'phossy jaw'.

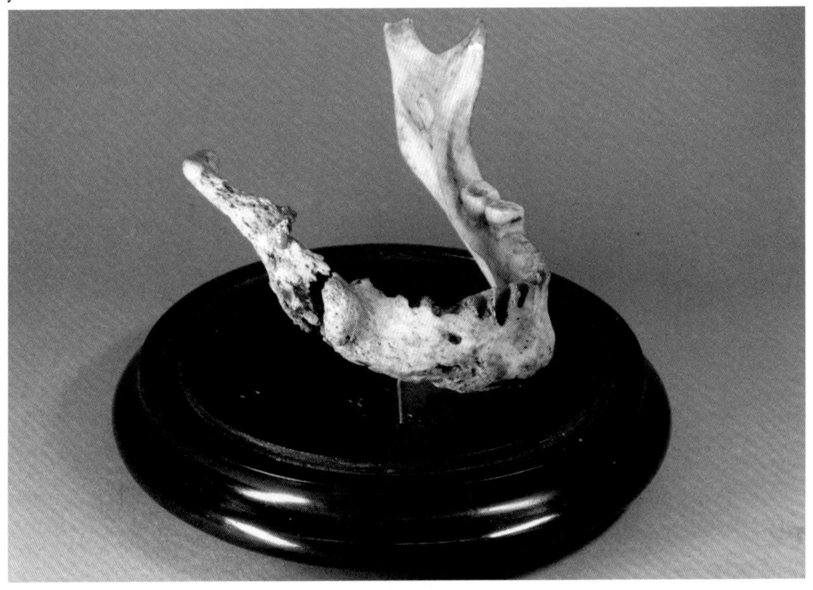

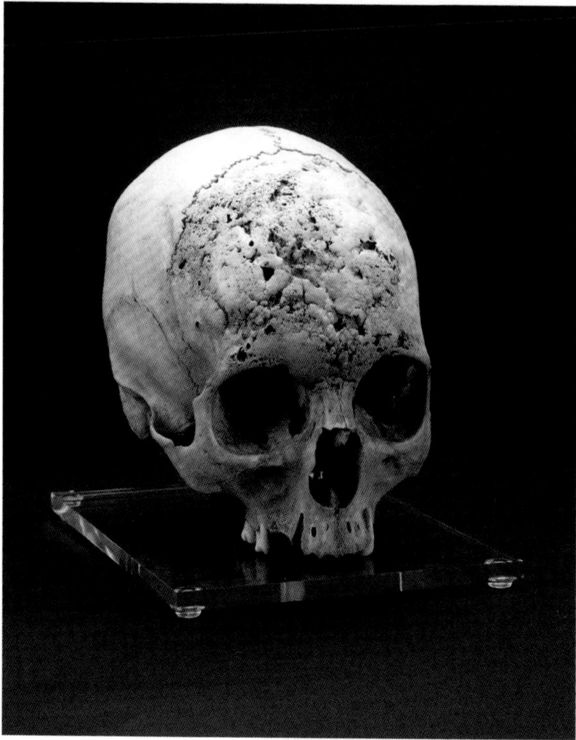

Left: Caries sicca, the 'dried-chewing-gum' appearance of the skull bones, is typical of syphilis infection.

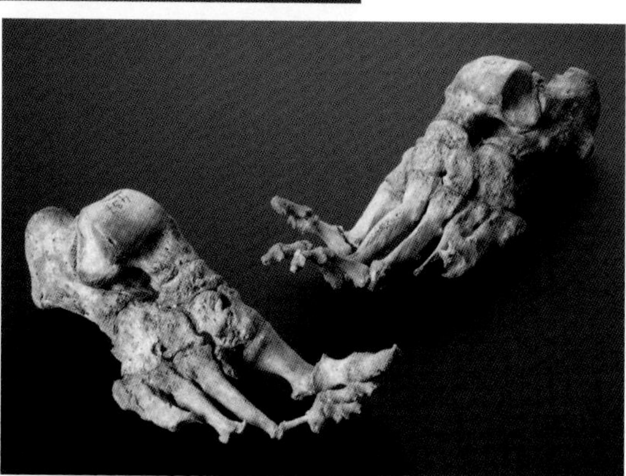

Right: *Mycobacterium leprae* infection causes paralysis and infection of the limbs, resulting in necrosis and loss of bone, as seen in these feet from a leprosy sufferer.

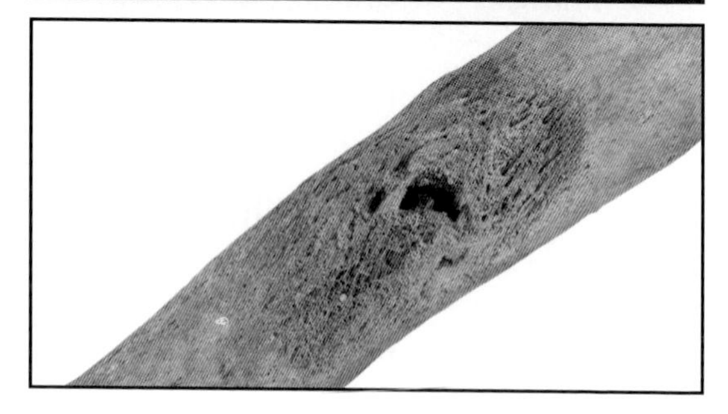

Right: Infection of the bone, osteomyelitis, is characterised by reactive bone at the site of the infection, as can be seen here.

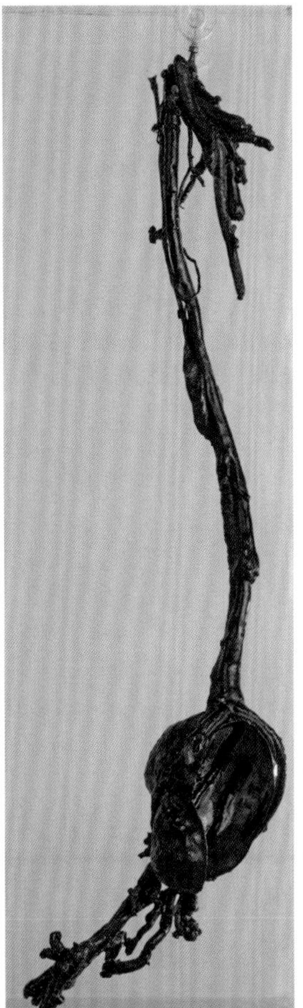

Left: The very first ligature of a popliteal aneurysm, carried out in 1785 by pioneering surgeon John Hunter on a 45-year-old coachman. The aneurysm may have been caused by vessels in the knee, already weakened by syphilis, being repeatedly banged against a coach box.

Below: Hutchinson's incisors. The moon-shaped notches in the biting surface of the teeth may indicate that this child was born with congenital syphilis.

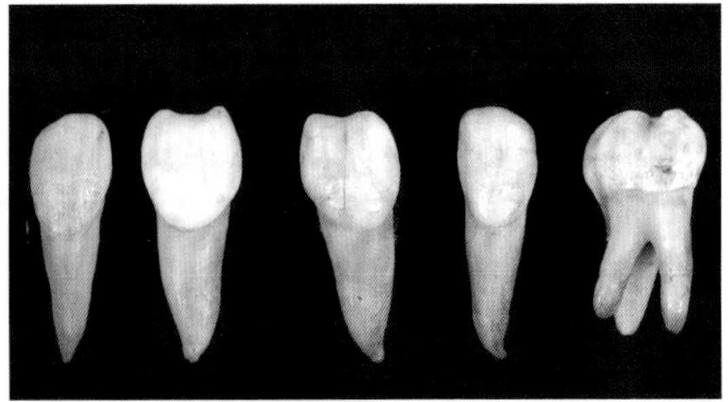

Left: One of the worst traffic jams in London history was set off by Handel's Music for the Royal Fireworks. It caused traffic to stand still and the 'fireworks' part of the programme set fire to the venue.

Above: Pollution, crowding and poverty in early modern London were among the reasons for the spread of disease. Here, London is imagined as 'A Court For King Cholera'.

Below: The filthy and unhygienic nature of the early modern city was palpable to contemporary Londoners in the air, and was common motif in period illustrations.

not an empire as such but a group of obviously networked sites occupying coastal Peru in the period before the Inca Empire, seem to have emphasised their military fierceness through the rather terrifying iconography of their ceramics and statues; these include scenes of human sacrifices as well as torture like disembowelment. The coastal cultures of Peru seem to have had some evidence of human sacrifice well before the professionalisation of the practice by the Inca, as testified by the incredibly sad evidence of a small infant buried alive as a possible foundation offering found at the ritual centre of Pachacamac. Far to the north, we have the Maya sinkholes where, beneath the water's surface, hundreds of bodies that had been ritually cut, dismembered and probably deliberately killed have been found. While there is no real agreement on the exact nature of human sacrifice in Maya culture, their elaborate culture of personal physical sacrifice* did culminate in practices that stacked up bodies down the cenote in the end.

More generally, the empires of Mesoamerica contained arenas for the sorts of violence we know from states all over the world. The Inca left their sacrificial victims in the cold dry air of the Andes, where they are found perfectly mummified today. Forensic testing reveals the narcotics they took with their last meals, as well as genetic and isotopic signatures of their relationships to the land and its occupants. The Aztec, who get quite a bad rap from judgemental European observers,† may have left piles of decapitated victims alongside the fancy folk in elaborate tombs, be on record plucking beating hearts out of chests, have raided the nearby non-Aztec populations to

---

* Critical to Maya ritual were practices like piercing the penis with a string of thorns and indulging in hallucinogenic enemas. One might pity the poor Victorian archaeologists who came across the evidence of all this and were forced to come up with *something* to label the enema tubes as in the catalogue of finds. Lesson being: never trust a Maya 'fan handle'.

† Decapitation sounds nicer than what the Inquisition would do to you. Just saying.

beef up their human sacrifice counts, and once killed 4,000 people for a ribbon-cutting ceremony.* But they also made these considerable sacrifices on behalf of their people – interpretations of murals on the sides of the great ball courts that form a centre point of Aztec cities suggest that on occasion the entire winning team of their version of basketball would be sacrificed,† which in many cultures today would no doubt be a profound act of abnegation.

What is it about urbanised states that makes humans into disposable – if valuable – property? This is a question that I'm very interested in, and I hope that my current field work in southeastern Turkey is going to provide some of the answers. The site I work at comes from a period of social revolution in Mesopotamia. There's *always* a revolution happening in Mesopotamia. Political stability does not seem to be a major feature of the region stretching from the Persian Gulf to the plains of Anatolia and the mountains of Iran. From the previously discussed development of agriculture to the resurgent insurgencies of the modern day, there is considerable backstory to the social and political upheavals that have wound up and down the length of the two great rivers that, in the words of the antiquarians, formed the 'cradle of civilisation'. In the summer and autumn of 2015, I had the dubious privilege of experiencing two of these tremors in political instability: the first being the resurgent campaigns by the Turkish state against the banned Kurdish resistance party (PKK), and the second being the collapse of the world's very first imperial-style network, the city-state of Uruk and its chain of colonial outposts wending up the Tigris and Euphrates all the way to the Basur River in Southeastern

* Based on historical accounts, this was the number of human sacrifices required for the re-dedication of the Templo Mayor in Mexico City.

† Imagine basketball where you're allowed to run on the walls, the ball can't touch the ground, you can't use your hands and the ball is 18 kilos (40 pounds) of solid rubber. Also, you die if you win.

Turkey. While the former was ill-advised from a personal security perspective and led to an unlooked-for skill in distinguishing the distinct reports of different gauges of weapon, the latter took place 5,000 years and about 2 metres (7 feet) down.

In July of 2015 I set out with Sarah Livesey, my research assistant, to the otherwise nondescript bottom-right corner of Turkey: Siirt Province. A colleague, archaeologist Veysel Apaydin, had put me in contact with Haluk Salğlamtimur of Ege University, who was running the vast mission of excavating the archaeology threatened by the Ilisu Dam Project. In an endless series of miracle reprieves, excavations of the river valleys leading off the Tigris and into the mineral- and metal-rich northern Mesopotamian hinterlands has been ongoing for nearly a decade. It was only when a particularly radical find came up at the site of Başur Höyük, just outside the city of Siirt, that a physical anthropologist was called in. In July of 2015 that was me, frantically sandwiching planning an entirely new research project in between visa appointments and research trips to remote Greek islands.

Sarah and I arrived on one of two daily flights into Siirt Airport, hungover after a brief recon in Istanbul, bags packed full of imported gin (her) and raspberry vodka (me). We came in over a landscape indescribably alien to someone used to the broad plains of America or the gentle green undulations of continental Europe. The river valleys of Siirt in summer are endless brown, desiccated hills, so many that they look like a bed sheet that's been scrunched together in the middle of a continent. There are mountains to the north and east, but the regular ripple of the hills, with no trees, no flats, nothing to break the monotony, is a pretty unique view. Once the little twin-engine bounced to a halt on the one-lane runway, we followed the crowd out into the single-roomed airport building, eventually emerging into the full wattage of a 45°C (113°F) Siirt summer day. Waiting to greet us were Haluk and the indispensable Man In Charge Of Things,

Nescat.[*] We quickly built up Team Anthro by poaching the more experienced – or possibly just masochistic – students and excavators, and with the rather crucial addition of Nil Muhafel as both assistant and translator. Thus began our adventure with the project I still insist on referring to as 'CSI Mesopotamia' despite the obvious dangers of copyright infringement – an incredible series of burials from the Early Bronze Age, dating back to 3100–2700 BC.

To understand the mystery we found under a good few metres of accumulated soil and debris, and why it was every bit as revolutionary as the roadblocks, gas, bombs and inescapable military presence of the modern day, we need to go back to the fourth millennium BC. In the fourth millennium BC, the world has no pyramids. There is no real written language. The most that can be said for the vaunted Urban Revolution promised by Childe is that in the several thousand years between the Neolithic and the Bronze Age, we find scattered across the world small individual 'villages' like the walled, defensible habitations of Jericho or Banpo in China. Some of these sites might be quite big and hold a lot of people, but they're relatively isolated, trading sporadically or with just a handful of external partners, which we can see through their unique repertoires of material goods and the limited number of exotic items their attenuated networks bring in.

In the fourth millennium BC, far to the south and almost at the mouth of the Persian Gulf, the settlement of Uruk kicked off the age of empires with a series of ugly, cheaply made bowls in a handful of standardised sizes.[†] The Uruk colonies

---

[*] Nescat's roles included, but were not limited to: driver of cars, procurer of ice creams, father to the adorable small girls who once ran 3k with me in pink flip-flops, owner of bunny rabbits with a distressing fondness for sleeping in engine blocks and foreigner-retrieval manager from the shopping malls of Siirt.

[†] That these unprepossessing bowls are considered one of the most exciting finds in the history of the discipline goes a long way to explaining why archaeologists are frequently disappointing conversationalists.

seem to have functioned basically as airport Starbucks for the industrious trade networks that funnelled raw materials into the rapidly growing city of Uruk. The standardisation of material culture found outside the actual heartland has always seemed to me to mirror the bizarre experience of walking into a familiar place in countries I have only just arrived in for the first time. We can see from the material culture at Başur Höyük that the time of Uruk came and went. But in the brief hiatus between empires, our site flourished, and the fancy tombs mentioned in Chapter 5 appear. In 2014, at the foot of the largest tomb so far uncovered, the archaeological team made a remarkable find. Several bodies, liberally coated in incredibly exotic beaded belts and necklaces, had been buried just outside a tomb door. The tomb itself was sealed to them, as it had been for more than 5,000 years, but there they lay, entwined and interred on the same level as the main burial. We know that at the roughly contemporary site of Arslantepe, far to the north and west in Anatolia, the very first signs of the practice of human sacrifice in the ancient Near East had been found in another stone tomb, this one decorated with partial teenagers in fancy hats. Just 300 years after our site, the Royal Cemetery at Ur would utterly exceed expectations and produce prolific mass burials that seem to be at least partially (if not entirely) the result of human sacrifice. The incontrovertible hundreds of human sacrifices that surround the early Egyptian pharaohs in their pyramids follow soon after.

As the project's physical anthropologist, my job is to work out whether the bodies lying outside the fancy tombs are victims of human sacrifice by carefully examining the skeletal remains for signs of cuts, punctures, strangulation or other methods of killing; to examine the excavation record for signs that the skeletons might have been bound or otherwise arranged in such a way as to suggest that they ended up there as a job lot of sacrificial victims. As an archaeologist, my job is to question what the sudden arrival of human sacrifice in the record of human achievements means. If these are really human sacrifices, we have to ask, 'Why here, why now?' If we want to understand the ultimate embodiment of structural violence, we must ask what it was about the political, social and cultural situation at Başur led to the development of human sacrifice. There are many competing theories; an analysis of human sacrifice in 'Austronesia' (the bit of the world between the Asian continent and the Australian one) has argued that human sacrifice is in fact a driving engine in maintaining social stratification. Joseph Watts, the lead author, argues that human sacrifice in Austronesia was a way of exerting social control, to terrorise or punish in a manner akin to the judicial killing we discussed above. The power of a 'special' group of people, an elite, to kill others is legitimised and codified through the act of human sacrifice. Peter Turchin, on the other hand, vehemently disagrees, and points to sacrifices that can't be interpreted as punitive, such as the gentle poisoning of the Inca children whose mummified bodies were found high up in the mountains of Peru. His argument is that if human sacrifice is such a terrific idea for stabilising your new hierarchy, everyone would be doing it; indeed, they would *still* be doing it. But human sacrifice has its big moment at the dawn of complex states, just like at Başur, and then fades away.

At the risk of sounding incredibly reductive, there is of course a third option: ritual. To hark back to the longstanding joke in archaeology that anything that cannot easily be identified must be 'ritual', and as such can be left alone while we get on with the things we do understand.* But ritual can be read another way: as a capacity to display power, which in the tenuous world of early elite hierarchies might have been *the* difference between having power and not having power. Shifts in regional complexity, in the way populations and networks of populations are organised, call for new and more impressive means of display, and the rituals to fit. The concept of human sacrifice as the logical extent of hierarchical societies' need for bling might be unpleasant, but then, so are humans in aggregate. There are many contexts in which

* *e.g.* What's Stonehenge for? Ritual! How was it made? Oh wait, I have detailed scientific explanations.

human sacrifice has been practised in the past, but in order to understand the wide range of potential human approaches to sacrifice, we need more specific evidence. This is something I hope to address in the near future, when I return to Turkey to analyse the remains more closely.*

Violence does not have to take the rather direct, loosely equal form of a teenage fistfight, or even a bit of conflict management as in the case of the Chumash. In the previous chapter, we looked at violence between (more or less) two people, *mano a mano*, spatula to cranium. In this chapter, however, we've seen that it's entirely possible to be at the mercy of not just *a* person, but in the sights of the rather more threatening *people*. Violence can and does regularly occur against single persons or subgroups within society, perpetrated by some socially recognised arbiter – a king, a priest, a witchfinder general – of moral or spiritual good.

The very first examples of both of these are only first observed, in whatever region they arise, with the dawn of complex urban polities. With the exception of the structural violence enacted in the domestic sphere on women and children, all of the other cases discussed in this chapter are possible only with a level of authority that comes with complex polities. Archaeologist Haagen Klaus argues that signs of violence, and of health disparities, can be interpreted as signs of structural violence (albeit carefully). He associates this with the presence of 'rigid hierarchies' and social complexity, both traits frequently attributed to urban life. Whether the prevailing view of cities as the forges of violence can be challenged on a purely statistical scale will have to depend on future excavation, but it's interesting that discoveries in recent years have pushed back the start dates for evidence of mass killing and fatal conflict. One might begin to wonder, if it were easier to find the dead of small-scale societies so

* Assuming that the state of emergency has been lifted; having probably tempted luck sufficiently during the 2015 season, there seems no real reason to do so again until the situation is slightly more regular.

inconveniently lacking the sign-posting of an enormous urban ruin, would we find that the experience of violence was closer to the higher levels observed ethnographically in small bands and tribal groups? What might be true for interpersonal violence and community-on-community conflict might be very different in the case of structural violence. While there are women and children in all societies, and their abuse may not be fatal, human sacrifice is a fairly profligate use of human life. It takes a large society, with unsubtle obligations and the hierarchies that enforce them, to have the power to coerce their own citizens (or those non-citizens unfortunate enough to be caught nearby) into death.

CHAPTER EIGHT

War! What Is It Good For?

We have looked at the skeletal evidence for injury in Chapters 6 and 7, but only in the context of interpersonal violence. This chapter moves from ice pick to pike, from the singular to the communal – when violence ceases to be about a single person or a small group of people and becomes a society-wide conflagration. There is certainly more than enough evidence to go around proving that humans are red in tooth and claw, but can we confidently say that war itself is a 'disease of civilisation', or linked to the rise of cities? Archaeology suffers from trends in the same way that all human creations do, and the idea of *Homo sapiens* as the weapon-wielding ape has a long antiquity and a great deal of cultural cachet. In the iconic opening scene of Stanley Kubrick's *2001: A Space Odyssey,* it's the image of an ape brandishing a femur as a weapon that presages our species' descent into violence. Were alien ethnographers to drop from the sky into the modern world, they would be confronted by the constant material culture of militarism: generals and admirals immortalised in bronze, triumphal arches and ritual tombs, or museums full of gaudy-coloured battle scenes and racks and racks of rusting swords.* But these are the memorials and material culture of a world fully enmeshed in nation-state politics. What about the time before? When there were no cities, with their complicated hierarchies and endless capacity for consuming the labour and resources of the territory around them?

Until the mass casualties of the world wars of the first part of the twentieth century, 'savages' were thought to be just that – savage. In the eighteenth and nineteenth centuries, when people were conveniently writing in

---

* And heaven help them if they land in North Korea.

European languages that still get taught in schools about the nature of war, the Enlightenment had by and large delivered. Civilisation was reaching an apogee from which it could never be torn down, so, as you can imagine, the twentieth century came as something of a shock to the system. And in the aftermath of truly global war, a tendency to look at the past through rosy glasses emerged in archaeological theory. A quick trip through of the fluffier bits of the internet reveals that this was not a phenomenon restricted to the social sciences, but a larger cultural meme. According to a certain school of thought, popular in many California parks and university campuses in the 1960s, there existed before the world of cities* a sort of perfect state of child-like innocence, as typified by the imagined lives of pre-colonial Native Americans. The inconvenient social complexity and capacity for ecological destruction of these actual groups of actual humans got lost in a dreamy, infantilising narrative, full of praise for a pacifistic wisdom and perceived harmony of indigenous existence prior to European contact. While many Native American groups in the modern era did and do hold pacifist ideals, such as the Hopi, who registered their conscientious objection to the US war in Vietnam, many did not; the evidence of the previous two chapters would seem to put this hoary old chestnut to rest.

Violence was a feature of life in the New World just as it was in the Old; in Asia as in Africa as in Australia, on islands, on continents, up mountains and down valleys. The violence under discussion so far, however, has been limited to the violence done against individuals, or, at most, small groups from within a larger pool. It's probable that not even the floweriest of flower children would deny the tendency of humans to resort to violence when the chips are down. In throwing aside the regimented chains of thousands of years of Abrahamic worldviews and throwing open those doors of

* And the draft, and polyester, and capitalism, man.

perception, the hippies had more or less reinvented the idea of a biblical Garden of Eden.* In more academic terms, the concept of the world before states, cities and complex societies – in other words, up until the end of the Neolithic and for many areas, well into the Bronze Age – is imagined as the last peaceful moment in our species' tumultuous history. The western cultures that gave us hippies also gave us radical reinterpretations of the past, and led by the much-maligned† archaeologist and ethnographer Marija Gimbutas in her work on the Neolithic, imagined a Europe before the Bronze Age ruled over by a peaceful matriarchal society. The Mother Goddess myth struck a chord with many, particularly during the cultural upheavals of the Vietnam War. Decades later, dozens of women – largely affluent, middle-aged and American – made the pilgrimage to supposed sacred sites of this overarching matriarchal power, even if their celebrations of the Mother Goddess raised a few eyebrows‡ and caused no little amount of cultural confusion.§ However, it wasn't just the consciousness-raisers of the Berkeley area who imagined the Neolithic as a world before war. It's worth asking: can you even have war before you have a big complex state, topped by cities, where kings and bureaucrats can command massive standing armies? Life before the creation of large centralised states – the Romans, the Aztecs, the Han – in

---

* Indeed, there is a school of thought that interprets Eve's apple as a withering indictment of agriculture, and another that reads it as a paean to pest control; the Garden of Eden is the advanced yogi of metaphors.

† Much of this maligning has the faint whiff of misogyny about it. Many of the now-disproven but at the time paradigm-challenging works by female academics of this period received such an abundance of derision that it has obscured their real contributions; see for example Elaine Morgan of 'Aquatic Ape' fame or the early response to Jane Jacobs' groundbreaking work on cities.

‡ Read: topless dancing around the Çatalhöyük mound.

§ Read: the house they rented in the very conservative nearby Turkish village mysteriously burned down.

what might loosely[*] be considered the Metal Ages was a thing of villages and regional exchange networks. Does war only take a village?

It may or may not help to define what is meant by 'war' in this discussion, even though it's a very common word and presumably easily intelligible to any reader with a vocabulary beyond the level of a five-year-old.[†] Given that this is a book largely about the past, you will probably also have an accompanying set of mental images, specific to your personal and cultural tastes, perhaps involving medieval knights on horseback, Roman legions or Japanese samurai.[‡] A standard definition would see war as 'armed conflict', but that doesn't quite capture the role of war in the lives of cities, states and cultures. The eighteenth-to-nineteenth-century Prussian military theorist Carl Philipp Gottfried von Clausewitz famously said that war is 'a continuation of politics[§] by other means.' This is taken to emphasise that war is a political act, by political actors, intent on political outcomes: power over territory, population or other resources. The French social theorist Michel Foucault retorted, some years too late for Clausewitz to benefit from it, that rather, it's politics that is the continuation of war by other means. This emphasis on politics belies the worlds that both men occupied – big-state societies with big-state actors, writing in a time where the concept of war had been very much codified into the practice we recognise from newsfeeds today. An anthropological consideration of war, however, is a much broader continuum

[*] Hey, if you wanted a simple narrative you should have gone with a less complex planet. Blame the Americas for making jewellery and not weapons out of their available metals.

[†] Special message to any readers who do not meet this criterion: well done making it past the Marxism jokes.

[‡] It is the author's contention that a significant subset of the readership will also include dwarves, elves, orcs, trolls, dragons, hobbits and wizards in this exercise.

[§] Also translated as 'policy'. Translation is difficult.

of conflict, ranging from the territorial ragings of primates,* through the many different modes of community conflict present in hunter-gatherer groups and smaller-scale societies, to the urban worlds Clausewitz and Foucault were considering.

In the mid-1990s Lawrence Keeley published a book that was determined to shake up the notion of a peaceful past among anthropologists, to the same extent that investigating it had shaken him up as a researcher. In the preface to his volume *War Before Civilization: The Myth of the Peaceful Savage*, Keeley explains his paradigm shift as the result of personal failure: having gone along with the prevalent narrative of a peaceful past for so long, the revelation that the site he was digging was not so much 'enclosed' as it was 'fortified' came as a considerable surprise. Fortifications imply the presence of something to be fortified against, and cows are not known for their siege tactics. Keeley divided anthropological theorists into Neo-Hobbesian or Neo-Rousseauian camps, with the former seeing war as endemic to the 'primitive' condition and the latter arguing the opposite. Both, Keeley argues, treat this kind of community-wide violence as a binary, with 'primitive conflict' on the one side of the emergence of states and 'war' on the other. In order to accommodate the history of human conflict in the history of human cities, we'd need more flexibility than even the religious symbolism of the Bible can provide.

Our species' fondness for massacres is of sobering antiquity, and cannot be simply attributed to the rise of cities and the social complexity they embody. Given our long history of violence, it's worth considering the developmental history of our genocidal – or at least communicidal – tendencies. We know that even our primate relatives the chimpanzees indulge in fatal raids on competing neighbours, and many researchers have argued that the history of human conflict is older even

* Special mention awarded to the determinedly homicidal chimpanzees here.

than *Homo sapiens* themselves, as the evidence from Sima de los Huesos would suggest.

In the complex cave systems underlying those same Atapuerca Mountains where our potential first-known murder victim was found, many more archaeological remains of Europeans from the last million years have been unearthed, including potentially even earlier victims. Far below the surface level at the site of Gran Dolina, excavations uncovered a spectacular find: the remains of 12 individuals dating back to around 850,000 years ago, with distinctive skulls and teeth that led to their description as *Homo antecessor*. And what remains they were: young men, women and children – mostly children – whose bones had been picked clean. Bioarchaeologists make an effort at distinguishing the normal wear and tear affecting bones that have been buried underground, tumbled through water, leached by acidic soil or otherwise degraded. The study of bone taphonomy is exactly what allowed researchers to identify that these bones were not just artfully tidy – they had been deliberately stripped of meat. From microscopic cut marks that match the stone tool technology of the time to the smashed bits of long bone where all the best marrow could be had, the bones of this hapless lot were treated indistinguishably from the bones of other hominin kills. This was cannibalism, embedded deep in the human story.

Anthropologists once again turned to the demography of the dead for answers, arguing that the 'kill profile' of the cannibalised remains didn't reflect a total massacre of some threatening outgroup, or any of the patterns associated with cannibalism as it has been practised by recent human groups. Instead, they theorised that the preponderance of young in the assemblage was more similar to the kill profile of chimpanzees, which target the very young of nearby groups in situations where territory overlaps and a resource is contested. The cannibalised remains from Atapuerca may be telling a story of rival groups, fighting over access to resources and territory. If they were deliberately killed, and not just eaten, then they may push back our evidence of homicide another few hundred thousand years.

To find out if war – true community-versus-community conflict – existed before cities, we have to examine very carefully two branches of evidence. The first port of call for the archaeological evidence of conflict is the dead. In Chapter 6 we discussed the forensic signs of violence; but here we are taking things up a notch to full-on conflict mode. What is there to distinguish murder from mayhem? How do we recognise the evidence of war? It's harder than you might think – there are no cenotaphs, no memorial fountains with inscribed names for the vast majority of human history. However, if there is one commodity that war produces in vast numbers, it's dead bodies. And not just any bodies – bodies with stab wounds, bullet holes, bits blown off, smashed skulls, fractured bones and a hundred horrible things besides. Obviously, war between nations of millions of men armed with high-velocity projectiles looks very different than between groups that number in the hundreds waving clubs. The evidence for both, however, comes down to the same principle: wounded warriors and mass casualties.

The difficult part for the archaeologist is that these pieces of evidence may exist in very different circumstances. Take, for example, a cemetery of 100 burials. A small portion of these burials contains weapons, clear indications of violent symbology, if not actual practice; these are akin to the 'warrior graves' that John Robb was talking about in Chapter 6. Do we accord the individuals thus buried the status of 'warrior' and assume that the prevalence of weapons is a straightforward measure of how warlike their society was? The classic interpretation of elaborate burials outfitted with all the tools of war was proposed by Paul Treherne in his article 'The Warrior's Beauty': the ritual treatment of the material culture of war implies that conflict itself is highly ritualised. Or perhaps we have a different case – a mass grave. If this mass grave contains the bodies of only men of fighting age, can we assure ourselves that this is clear-cut evidence of conflict, because a normal society is not made up exclusively of young men? It's at this point that the evidence of bioarchaeology becomes crucial. Weapons do not necessarily a warrior make, and men don't always die on the battlefield. A cavalry pike through the head, however, is pretty unambiguous, and a pit of 20 skulls with pike wounds, more or less unarguable.

This is the evidence we begin to find scattered through the archaeological record. Much of the earliest period, before sedentary lifestyles come in with the Neolithic, is sparsely represented; whether because we haven't found the Mesolithic dead yet or because there just weren't many Mesolithic people around remains unsatisfactorily answered, given that what archaeologists *do* find tends to be either in areas of modern habitation or unsystematically encountered. Even as we come forward in time, there are very few examples of obvious conflict. With notable exceptions, isolated finds of traumatic violence in the past, like Ötzi, cannot be directly interpreted as evidence of war – they might as easily be signs of interpersonal violence. The notable exceptions are, however, very notable, and they stretch far back into our hunter-gatherer past.

In January of 2016 a truly shocking story appeared in the pages of *Nature*,* upending our definitions of war. Nearly 10,000 years ago, on the western shores of Lake Turkana in modern-day Kenya, a small group of foragers – men, women and children – were slaughtered with a combination of arrows, blades and blows. The remains of 27 individuals† were discovered at the site of Nataruk, haphazardly discarded into the prehistoric lagoon. A team led by anthropologist Marta Mirazon Lahr found that many of the skeletons bore clear evidence of considerable violence: embedded arrowheads,

---

* Official religious text of the modern scientist.

† Depending on your point of view; the grim discovery of a foetus *in situ* in the abdominal cavity of one of the women might make 28.

smashed skulls and fractures to the ribs, spine, knees and hands. This was a massacre, a complete eradication of a community of foragers, committed by the only other possible suspects: another group of hunter-gatherers.

A millenia or two later, in Bavaria, the cave site of Ofnet seems to offer a slightly different form of evidence: nests of human skulls. The Mesolithic cave would appear to have been embellished with rather morbid 'conversation pieces'. The skulls (and just the skulls) of around 34 individuals had been interred in circular pits near the entrance. The majority of the individuals were quite young: 46 per cent were under six years old, and there were more females then males represented. Nine of the heads had clearly been decapitated by hands human enough to hold a blade: the very top vertebrae of the spinal columns were still attached, and the marks where the heads had been severed from the rest of the body were clearly visible on them. The original analysis of the bones in the 1930s found evidence of slashing wounds to the skulls and considerable trauma to the back of the head: the rear of the skulls had been caved in. Thirty heads in a pit, cut up and stove in, seems like pretty damning evidence that Mesolithic Bavaria might have been considerably less welcoming than the modern tourist authorities would have you believe. The damage to the skulls of Ofnet Cave, however, isn't the smoking spearpoint that archaeologists originally hoped for. Arguments have been made, on re-examination of the archaeology and further radiocarbon dating, that some of the material could have been deposited at different times – and so are not evidence of a battle with mass causalities. That still leaves the five adult individuals in the skull nests that have clear evidence of fatal blows from a 'chopper'-like tool with considerable explaining to do, but no incontrovertible proof of war.

What evidence we do have of war from the time before cities (and states) seems to fit largely well with the expectations of raiding and conflict over resources that we know from ethnographic literature, where fatalities occur but are not necessarily extended to the entirety of a community; nor are they restricted to just one group of 'warriors' – for example, the fallen include women and children as well.* However, we are increasingly elaborating on the range of violence open to people without the benefit of cities. The Bronze Age of Northern Europe, for instance, is devoid of cities but certainly not of warriors, and new evidence has conclusively demonstrated that these warriors did, in fact, go to war. The Tollensee River in Germany is the location of an impressive recent discovery of some hundred armed men (and five very unlucky horses) gone back to the mud shot full of arrows. These warriors may not have come from a city, but they certainly died for a political entity large and complex enough to compel them to fight, and not just the once: the weapons they bear are bronze and impressive, and their bones show evidence of earlier trauma.

Contesting territory, violently and with fatal consequences, is nothing new. The dawn of agriculture and settled life in many parts of the world sees, if anything, an upswing in community-wide markers of violence. Robb's work in Italy is not an isolated case. In the canyons of modern-day New Mexico and Colorado in the Southwestern United States, archaeologists have revealed hundreds of bodies under the desiccated remains of burned-out pueblo houses. Bodies of men, women and children show signs of a violent end – cranial trauma, unhealed fractures – before being unceremoniously deposited in a house that was then burned to the ground. Of the 23 bodies found in the collapse debris at the Ancestral Puebloan† site of Sand Canyon in Colorado, nine had fracture wounds that clearly pointed to violent deaths, and all of those nine had skull fractures. Other sites in the region show similar patterns, leading archaeologists to suggest that the environmental

* This is not to denigrate women's offensive or defensive capabilities, or to make blanket assumptions about social roles in the past; it is simply less likely that a mixed-age and mixed-sex group would share that specific a social role. I would argue that a society dependent on a child warrior caste is not going to succeed in the long term.

† A culture perhaps better known by the name the Navajo gave them: the Anasazi.

changes occurring in the thirteenth century AD caused sufficient stress in the dispersed Puebloan villages that fatal cross-village raids in these Neolithic cultures for dwindling resources became a major danger. All across North America, where we have a number of groups who transitioned to agricultural life at different times and in different places, the evidence is mounting for a far more violent Neolithic than previously imagined – the bodies are quite literally piling up.

Massacres are, despite the mass casualties, still not quite the same thing as Clausewitz had in mind when he wrote about war. The dead of the Neolithic death pits, like the skulls in Ofnet before them, speak of violent villagers and not professional killers. The startling finds at Nataruk are so far the first conclusive evidence of *modern* humans massacring an entire community, but it's difficult to argue that a single incident marks the engagement we know as warfare. The Nataruk may have been defending territory, but they did not hold it, or control the people on it, in the way an urban state does. The Bronze Age villagers who died at Tollensee may well have been contesting territory, with some of the trappings of professional warriors, but they may not have been Tollensee-local at all. There is a suggestion that the fighters may have actually come from further south, where war was a more firmly professionalised occupation. In the end, it's a bit ridiculous to segregate professional war and continuous or opportunistic raiding as features occurring before or after the advent of some arbitrary civilisational achievement level; it's equally pointless to try to corral 'primitive' war into categories of effectiveness or organisation. Deck chairs have been rearranged to greater benefit. It may be more helpful to try to consider the difference between using violence in short-term acquisitive raids versus towards a goal of political subjugation or control. That's not to suggest that war, even in* modern nation-states, isn't about acquisition of something, but rather to say that the mechanisms for violently achieving those ends can be either opportunistic or built into the architecture of

* Especially in.

society. The difference between the world of villages in the Neolithic and the more complex urban centres of later periods comes down to evidence of not just malice aforethought, but the threat of malice as a constant thought.

The second piece of evidence for war in the past, which we have finally got to, is the one that set Keeley on the path to re-integrating violence into the human story. And what Keeley discovered wasn't actually even bodies, it was … ditches. Defensive ditches. The earthwork fortifications of hillforts have been interpreted in a variety of ways, with periods of militarist interpretation followed by alternative explanations emphasising cultural symbolism. Ditches lead to walls, and walls mean that you would very much like an 'inside' area to remain separate and distinct from an 'outside' area. It also suggests that you have reasons to be wary of things outside that want to get in (or vice versa). This is the sort of planned-for malice that seems to be a feature of the more complex human hubs of the world. It's the violence aforethought that explains the walls that spring up around villages in early agricultural phases from around the world – from the earthworks of the Longshan culture in China to the fortifying walls (and neat lines of accompanying sling stones) found at coastal Ostra in Peru.

The walls of Jericho, for instance, ring loud in Western ears because of their role in the Old Testament,* but the very first incarnation of Jericho's walls, uncovered by the painstakingly systematic work of Kathleen Kenyon, may have been fortifications against a rather more implacable enemy than marauding Israelites. Kenyon, perhaps one of the most fascinating archaeologists to have lived and worked in the twentieth century,† had something of a habit of bringing a

* And, for a very select few, because of the seminal experimental Dadaist industrial stylings of Cabaret Voltaire's song 'Walls of Jericho'.

† It's a crowded field. Dame Kathleen Kenyon's extraordinary person and achievements cannot possibly be covered here, but biographies exist. It is enough to know that she essentially pioneered the unique archaeological skill set of being able to fix a car, operate a camera and rock double denim.

ruthlessly systematic approach to the overly romanticised archaeological sites of the Holy Land. Despite a fondness for leaving enormous baulks of earth for no apparent reason at the neat corners of her trenches, her careful excavation work found that, under the much later Iron Age fortifications, a very early series of walls had encircled the site, topped by a stone-footed tower. Based on the pottery she hazarded that the foundations went back to the Early Bronze Age, a date later confirmed by radiocarbon testing. Careful study suggested that these walls were pitched in such a way as to protect Jericho from flooding, not assault by trumpet. The lookout tower was reinterpreted as a ceremonial hub, and the walls of Jericho fell again to the post-processual forces of archaeological theory.

So, why should it be that when we see a wall around an ancient site, our first thought should be that it was built to keep other people out? If Jericho's walls were built against the elements, why do we jump so easily to interpret *all* walls as defensive fortifications? For many archaeologists, even the most fortress-like settlements in Europe – big, mounded escarpments with earth ramparts collectively falling under the heading of 'hillforts' – can be seen as more than evidence of warring tribes. Much as temples and ritual centres were a collective project in the very earliest settlements, big walls could also be the ritual focus for group activity and general showing off – something to quite literally rally behind. If Jericho's walls were for keeping water out, and the hillforts of Iron Age Europe were for keeping cattle in, perhaps the appearance of city walls and other fortifications in human history cannot be immediately interpreted as evidence of increasing conflict.

Not everyone agrees with this. Keeley in particular takes umbrage at the interpretation of walls and earthworks as being symbolic – rather than defensive – in function. The Neolithic period in much of Europe is identified with evidence of the spread of the supposed peaceful communities of the Linearbandkeramik (LBK) cultures, whose distinctive ceramic styles creep into Europe alongside farming and domesticated animals. We've seen the progress of farming culture on the radiocarbon maps of the Neolithic Revolution and discussed the population shifts that may have accompanied this transition in Chapter 4 – a case of pots *and* people. Towards the tail end of the LBK period, around 7,000 years ago, the villages of these early farmers became *fortified* villages. And over the last few decades, we have begun to accrue evidence that the people of the Neolithic had very good reason for putting up walls.

In the 1980s the discovery of the Talheim 'death pit' in Germany was an international sensation. The publication in 1987 of the findings of the archaeological team took everything that had been thought about the sedate, sauntering progress of agriculture (and agriculturalists) through Europe and chucked it out the window. Thirty-four individuals were found bludgeoned to death with LBK chopping tools (axes and adzes) and cast willy-nilly into a pit, discarded human carcasses without normal ritual positioning or accompaniment of grave goods. Just as at Nataruk in Kenya thousands of years earlier, the dead included men, women and children in such a variety to suggest that they might have represented the whole of a village. The crushing injuries that killed them were dealt to the back of the head, suggesting that they had been taken by surprise or otherwise not resisted: there were none of the parry fractures or defensive cuts to the hands or arms that modern forensics would identify as evidence of fighting back. Not long after, nearly 640 kilometres (400 miles) due east, in Austria at the site of Schletz-Asparn near Vienna, a ditch that may have held as many as 300 bodies was partially excavated. Extrapolating from the 60-plus remains analysed, archaeologists identified a massacre nearly identical to the one at Talheim – death by Neolithic axe, and in one case, arrow. In 2015 yet another LBK massacre was discovered at the site of Schöneck-Kilianstädten, 145 kilometres (90 miles) due north of Talheim. Twenty-six people died from a combination of arrow and axe wounds. Their lower legs were smashed in, with a small majority of the individuals showing compound fractures to the tibia (the bone that makes up your shin) that occurred around the time of death.* In a final new twist, there appear to have been no younger adult females among the dead – they may have been taken captive.

So, we can definitively say that the world before cities had its massacres: community versus community in armed conflict. As populations expanded, so did the opportunities to rack up casualties. The mass deaths described in Germany, Kenya and Austria do not seem to adhere to the ethnographic examples we have from anthropological studies of small-scale warfare – raids that take captives, or at least leave a survivor or two. Massacres seem to occur in specific but variable circumstances: some combination of population pressure possibly accompanied by climate change, and all of a sudden the bottom drops out of your way of life. But, as the litany of depressed cranial fractures from Germany and Austria tell us, these massacres are carried out with the weapons of … agriculture. Adzes and axes are no swords. For all of the death they bring, these inter-village conflicts lack the professional militarism, the special social status, of organised, institutional warfare. Just because you don't have weapons to hand doesn't mean you can't have a genocide: the horror of Rwanda in 1994 was inflicted largely with agricultural tools. When we do see early evidence of conflict using proper weapons like those found at Tollensee, we can't quite trace back the origins of the warriors, let alone their war, so it's very hard to argue how integrated they are into the culture of the muddy river valley in which they were found.

Archaeological evidence of 'warriors' has been encountered far more often than the skeletal evidence of battle. This might be attributable to the phenomenon John Robb describes, where the prevalence of symbols of war found in burials – axes, shields, chariots, spear heads – does not map onto the evidence for rates of violence revealed by the skeletons within.

* One can only hope it was afterwards, but that seems overly charitable.

Keeley considers the interpretation of metal found in elite graves with fancy bronze weapons as merely a form a material wealth as somewhat missing the point. If these symbolic spearheads are commodities – money – then what about old Ötzi the Iceman, who died with a fully hafted example? Were, Keeley quips, his bow, dagger and arrows merely small change?

We now know what Keeley did not in 1996 – Ötzi died a violent death, and his spearhead seems incontrovertibly to be a working weapon. But just because one bronze spearhead is functional doesn't mean they all are. Out in the far southeast of Turkey, our ongoing* excavations at Başur Höyük have unearthed spectacular bronze finds. The cemetery that pops mysteriously into existence at the collapse of the first Uruk network holds dozens of elite graves. They are full to the brim of bronze objects: elaborately moulded staff toppers show animals and designs of particular sophistication, personal ornaments like pins, and, unsurprisingly, bronze spearheads. So far, so warrior grave. And they would have pulled off the image of gritty fighting men too, if it hadn't been for the fact that these spearheads were still in the Bronze Age equivalent of the bubble wrap they came in. Still packaged in delicate traces of fabric wrap ties, job lots of 75 spearheads, mint condition,† were carefully deposited alongside hundreds of ceramics, thousands of beads and whatever organic goods we have lost to time. After a brief bioarchaeological assessment of the individuals found with these amazing stockpiles of weapons, it seems unlikely that, for instance, the 12-year-old child found among the other bodies in the most elaborate grave would have wielded quite such an arsenal in life.

We know that many of the early urban centres of the world heralded military conquests with elaborate monuments and

* Given the political situation in the current moment, this is a statement more of optimism than fact.

† Some wear and tear from 5,000 years under the earth, as might be expected.

stelae – written and iconographic celebrations of victorious battle that were integral to their view of themselves.

The famous Stele of the Vultures, currently housed in the Louvre, is such an object, left as a permanent testimony to one of the earliest known organised battles in history, sometime around 2450 BC. It tells of a military victory by Eannatum, leader of the Sumerian city-state of Lagash. The dual carved stone faces contain numerous registers of scenes of battle and eventual victory, but the one that always sticks in my mind is the mound of naked corpses that accompanies the text below, which glorifies in slightly ritualistic tone the 'multitude* of corpses' that will 'reach the base of heaven'. Lagash itself is one of the very first of the Mesopotamian city-states that appear in the mid-third millennium BC, just after our cemetery up north. The war described on the Stele of the Vultures is the closest thing we have to evidence of the first city-on-city conflict; this is another thing altogether from the village raids or community violence we have discussed previously. If the multitudes described on the Stele are literal, and we take an acceptable casualty level of about one in five (a guestimate frequently employed in premodern warfare), then it implies that the actual size of the enemy army was some 18,000 people. A more reasonable estimate comes from the actual written account of tablets found in Shuruppak dating to around 2600 BC, which notes the number of warriors that the king financed. This earliest of standing armies, it seems, may have only been 600 to 700 strong. Either way, it's clear that by the time of the Mesopotamian cities, war was an institutionalised practice.

It's no great surprise that the professionalisation of war through the sponsorship of warriors supported by a hierarchical urban city or city-state leads to the sort of wars Hollywood spends so much money imagining. With the exception of a few modern nation-states, a considerable

* 'Multitude' in Sumerian is very specifically 3,600, for reasons that I have never really understood.

amount of illustrated matter aimed at young men, and the always excitingly different Scandiweigans,* the word 'warrior' has meant for the most part young-to-middling-aged men. While we discussed in Chapter 1 the hazards of doing demography on dead people, in the case of identifying evidence of war in the past, a pit full of young men with holes in their heads is a pretty incontrovertible bioarchaeological signal that conflict has occurred. What, however, is the reality of this manly conception of war, where it is men who fight and men who die? Is that what really changes, from Neolithic (and earlier) massacres that include all-comers, to the exclusively male carnage of the Somme? It may be that the killers in both cases were young men; it may be that young men in societies past and present, small and large, are frequently gathered in groups and societies for some sort of violent activity – both archaeology and ethnography tell us this. But it certainly cannot be said that the warfare that pits cities against cities or states against states does not result in the death of entire communities and the massacre of truly horrifying numbers of non-combatants – that was the lesson my fourth-grade reading of *Sadako and the Thousand Paper Cranes* was designed to teach. Can we say that warfare before cities was really that different to what came after, in terms of the people affected?

Cities are the harbingers of complex polities; they are defined by their complexity and are the necessary engines of the state systems that contest territory and power to truly horrific costs. The twentieth century AD saw death on a scale unimaginable (and quite literally impossible) in the twentieth century BC. Individual battles killed over a million people – by some estimates, the same number of humans alive on the planet when the whole Neolithic experiment started. If the mass casualties of the past century are an indication of the destructive power of warfare in an urbanised, hierarchical planet, then perhaps Charlton Heston was right and we really

* The portmanteau in which I include all of the countries you think of when I say 'Viking'.

are all maniacs, doomed to war our species to extinction and make way for the peace-loving apes to take pride of place on planet Earth.* It's very difficult indeed to imagine a world *more* prone to war than the one we currently inhabit.

And yet. We can think back to those symbolic warriors, going into their graves with less-dented skulls than their Neolithic counterparts thanks to the diversionary tactic of heavily regimented status hierarchies that make violent competition a bit of a dead end. Even the relentlessly self-promoted violence of the Moche culture of coastal Peru, with its dismemberment-themed pots and artistic depictions of bound and bleeding prisoners and human sacrifice – even they seem to see a fall in numbers of cranial fractures just as the violence in their artwork increases in the run-up to full urban-state society. We can desperately try to avoid picturing the smashed tibias of the men, women and children found in a shabby Neolithic mass grave thousands of miles from the nearest 'city'. Or we can follow Keeley's final argument: that it's civilisation that has pacified us, and cities that have saved us. He suggests that if you were to take even the worst excesses of twentieth-century warfare as a proportion of the people at risk of being killed in wars, the overall risk of death through some form of community-on-community violence is around 0.5 per cent. That works out (in Keeley's calculations) to more than 1,800,000,000 lives saved in the twentieth century alone, versus the deaths if warfare been conducted the old-fashioned non-urban way. While this seems an extraordinary supposition, a very late-breaking piece of research has confirmed that our species has very much lived up to the violence of our clade. In late September of 2016 a *Nature* paper described the background level of homicidal violence in mammals and primates – including us – at around 2 per cent, taking into account our social and territorial leanings, which bump you up on the mammal murder scale. Mammals cover a range of greater or lesser homicidal tendencies, and primates in general come out

* Obviously not. Clear evidence that you should never, ever believe anything Charlton Heston said.

at the dripping-with-blood end. Laying out the story of human killing *over time* against our primate relations, however, comes up with a pattern that Keeley would recognise. There is a rise in violent human-caused deaths from the Palaeolithic, when we more or less fit the expected mammalian trend, to the absolute peaks of the American continent's genocides in the period of contact – but a dramatic fall ever after. This collapse in the amount of lethal violence in our species coincides with exactly the set of circumstances Malthus gloomily predicted would be the worst for humans: higher population density. High population densities seem, in the human case, to have a pacifying effect.

By the time we come to armies, however, we have a new set of problems to deal with. Armies – proper, standing armies, not just the conscripted surplus population you can commandeer off the land – are dangerous for more than just the points on their arrows or the flaming balls of pitch in their catapults. Armies may be supposed to march on their stomachs, but the reality of military life in the past is far more like marching while emptying your entire gastric tract. Armies are full of disease, passed rapidly in sub-hygienic conditions between immune systems weakened by travel and combat. As an instrument of state power, they move between urban hubs, taking the diseases of one to the unlucky inhabitants of another, often bringing back entirely new diseases as a souvenir of the experience. At least the threat of armies comes at the speed of a slow march. If you think that urban empires with the population and power to wield 500,000 men in battle 2,500 years ago in the Persian campaigns into Greece was bad, wait till the war that we wage is against death that comes at the speed of a sneeze.

CHAPTER NINE

Under Pressure

The main preoccupation of the human species has been, and probably always will be, other people. Reproducing more of them, feeding them, controlling them, organising them, making them love and adore you – these are the things that all of our human achievements grasp at. The previous chapters have covered some of the ways in which our preoccupations with people have come back to haunt us in our recent past – essentially, the increasing density of humans on the planet, engaged in types of living that leave an ever-larger and -heavier footprint on the earth. The revolutionary torpor of the Neolithic, trudging for thousands of years across continents only to be reinvented somewhere thousands of miles away and begin again, left a legacy of people. Everywhere. As the people pile up, so do the places that hold them: villages become towns, towns become the very first cities. Across the planet, the concentration of humans led to a concentration of resources, often in the hands of a few.

We've looked in Chapter 5 at the physical stigmata of being left with too little in an increasingly unequal world. The last few chapters discussed our violent methods of recourse in problems we have with other people, whether it's in order to control them, to take their things or to inflict a more purely personal violence. In the next few chapters we will start to look at the final – and fatal – consequences of living in a densely populated urban world. Cities have a fairly bad reputation when it comes to coming up with ways to kill off their inhabitants, and one of the most difficult cases to argue in their favour is that of infectious disease. These chapters deal with the delicate balancing act our species has to perform – we want to be close to the centre of the action, but at the same time there are all these other people there, and some of them look like they're definitely sniffling. Cities tip the balance between the number of people potentially around to be infected and the viability of the infecting organism. Cities attract people, and people carry disease. But what if it's in cities that all those diseases circulate, mutate and adapt … to us? No virus wants to die alone; what if we were to make the argument that cities give infectious disease a chance to temper their malevolence, to become endemic conditions that do not kill the majority of their hosts? In this chapter we will look at the very peculiar case of the *Mycobacterium* family of infections, which offers a strange insight into the way cities have changed our experience of disease.

It's hard to imagine an animal species that doesn't come with its own particular pairing of infectious diseases – in recent years there has been considerable media coverage of swine flu, avian flu, badger-borne tuberculosis, mad cow disease and a number of others that alarm us with their potential to jump across to our species. We've already dealt with the diseases that our domesticated animals have given us in Chapter 3, but there are always other vectors for infection to be considered. I have indelible and slightly salt-crusted memories of my very first 'away' field school experience on the Channel Islands off the coast of California. After a thrilling ride on a coast guard boat that included racing dolphins, choking down sea spray *à la* Kate Winslet in *Titanic* and generally feeling smug about not getting seasick, we arrived on the island of Santa Rosa with our camping supplies for the weekend. The course coordinator, the highly respected archaeologist* Jeanne Arnold, had strict instructions. Most of them concerned not messing up the fascinating archaeology of the Chumash people who had occupied these islands, but the warning that stuck with me was the imprecation to wash

* And general inspiration for young students unsure of what they want out of life, who spend university break times sulking in their car listening to punk albums and chain smoking, until you encourage them out into the field – a graciousness I will never forget; I owe a huge debt of gratitude to Professor Arnold, whose name as you may notice appears quite a few times in these pages.

the tin cans our food* came in before opening them. The island mice, which scurried incontinently along the tops of the storage racks in the research station kitchen, were carriers of hantavirus. In my 20 years in California, I had never realised that a virulent killer disease was lurking just a few miles off the coast; the relatively rare hantaviruses can cause either respiratory distress or haemorrhagic fever, making them essentially a hipster version of Ebola. Mouse pee can kill you, prairie dogs are a reservoir for bubonic plague and armadillos carry leprosy; these are the sad facts of adorable mammal life.

For many researchers, the human story of infectious disease is a tale of two halves: an Edenic situation before the rise of sedentary farming life, and a quagmire of poxes and plagues ever after. This tipping point is known as the epidemiological transition, and traces of the increased impact of infectious diseases on our species really fluoresce in the archaeological record after this point. A new source of information, however, is challenging some of this orthodoxy: ancient DNA (aDNA) has started to give us a picture of the effect of infectious disease far further back in our evolutionary history than our trowels can find. This is less surprising than it may seem – only a handful of infections linger long enough in the human body to cause visible changes in the skeleton. Most have the decency to kill you quick, before your bones have a chance to react.

New advances in studying aDNA have shown us that Neanderthals and Denisovans, evolutionary (kissing) cousins who contribute a small portion of the genetic material to the modern humans whose recent ancestry lies outside of Africa, may have experienced their own epidemiological transition. According to a recent review by Charlotte Houldcraft and Simon Underdown, some of these cousins' adaptations to European diseases (like the encephalitis carried by local ticks) may have trickled into the modern human genome in Europe,

* I maintain to this day, in the face of considerable opposition, that SpaghettiOs are in fact food.

while a similar introgression may have occurred in people who eventually settled in Papua New Guinea, affecting genes associated with response to dengue, influenza and measles. We've traded infectious bacteria with a host of other hominins; there is now evidence to suggest we modern humans actually gave *Helicobacter pylori*, a largely asymptomatic infection that can occasionally be involved with more unfortunate gastrointestinal problems, to Neanderthals. It may be that further genetic work will uncover other transmissions across 'species' lines; perhaps even infections that the rather shallow Neanderthal gene pool was unable to survive. No wonder we gave them ulcers.

If we have always lived (or died) with infectious disease, why does it matter in the story of how we have built our modern urban world? Some researchers have suggested that, infectious disease was actually what kept us penned up in Africa for so much of our evolutionary history. Archaeologists Ofer Bar-Yosef and Anna Belfer-Cohen argue that disease must have been a major factor in keeping early numbers down, just as it is in chimpanzees today, and it wasn't until we escaped the all-you-can-infect buffet of local animal-borne diseases in Africa that we could get our numbers up. This may not be the most parsimonious explanation of hominid geographic expansion, but it does present an interesting idea: that diseases can determine our species' ability to expand into new territories and new ways of living. And of course, the converse is certainly true – as our species pushed into new landscapes and new lifestyles, we created new opportunities for disease. The epidemic transition coincides with the big lifestyle changes of the Neolithic because we introduced new disease risks and upped the likelihood of old ones. We started storing grain, which attracted a wonderland of commensal pests. Those pests are the very same suite of animals that show up in, say, Disney's *Cinderella*, but instead of bringing cartoon laughter and song to oppressed workers, rodents and birds brought bacteria in their poo, fleas in their fur and species-jumping viruses to crowded, dirty human habitations. Not only did we have new animals to trade infections with, but we spent a lot more time with our old ones: domesticating them in-house meant a much better chance of picking up their diseases. Even that hardy pastime of tilling the soil carried its own risks: parasitic worms and worse lurked in the dirt, just waiting to make the jump to human hosts.

Dirt, animals and people. These are ideal conditions for breeding disease. But this chapter is going to look specifically at the case of infectious disease caused by the *Mycobacterium* family of bacteria, which are very much diseases of people, by people and for people, though we're not averse to spreading the bacteria around a bit in either animals or dirt – more on that in a moment. We will start with absolutely no one's* favourite condition: leprosy.

To be a leper is to be untouchable: to die horrifically disfigured, shuffling off the mortal coil with your rotting bits leaving a trail of putrid wreckage behind you. There may or may not be indistinct and garbled moaning. The leper crab-walks through the popular imagination like a sort of Frankenstein's monster of every anxiety dream you've ever had: teeth falling out, fingers flopping off and probably some sort of medieval jeering in the background just for effect. They *sainted* Mother Theresa because she worked with lepers, they lionised Che Guevara for his unstinting commitment to delivering medical care to those most wretched souls,† and you wouldn't touch one with a bargepole. 'Be thou dead unto the world, but alive unto God' (*mundo mortuus sis, sed Deo vivas*) – these are not the words you want to hear from your doctor. And yet. Popular imagination has much to recommend it, but does it know what it's talking about when it comes to leprosy? Not so much, but that, it seems, is more than enough to structure the moral and spiritual treatment of leprosy through its long history in our species.

* I exclude bioarchaeologists, of course; we love a bit of leprosy.

† Fine, OK, also because of the sexy communist revolutionary stuff and quality hat-wearing.

What actually happens to a body infected with leprosy? If that body is relatively healthy, or at least possessed of a decent immune response, not much, for quite some time. Leprosy acts more aggressively in people with lowered immune status; in the modern world, cross-infections might cause this, but in the ancient world we might suspect the twin devils of poverty and misfortune. The most aggressive changes to the skeleton are seen in what is called lepromatous leprosy (*Mycobacterium leprae*): sufferers develop all of the symptoms mentioned below. Tubercular leprosy (*Mycobacterium tuberculosis*) is more of a disease affecting the skin, and while still a problem in the modern world, it's very difficult to detect archaeologically. Leprosy affects the nervous system, systematically destroying peripheral nerves in a process one medical text described as '*formatico*', which basically translates as the feeling of ants crawling on your skin. Nerve damage is not a fantastic thing for your extremities: it means loss of sensation and paralysis, which result in extra stumbles and extra cuts to the hands and feet. These can go unnoticed and untreated, especially in the farther-away feet, which is why it's actually secondary infections that are responsible for the rather horrifying tissue necrosis we associate with rotting lepers. The bones of the extremities show characteristic changes associated with infection response: loss of bone, changes to the articulations of the joints through inflammation and the body's attempt to mount a rear-guard action, and formation of new bone and new articulations as the body tries to adjust to a newly paralysed limb.

Large amounts of the infective bacteria itself seem to hover around the nose and mouth, swelling the soft tissue and causing weeping sores that lead to a diagnostic series of changes to the face: the very foremost part of the nasal bone is shrunk away, the gums around the front teeth are eroded along with the underlying maxillary bone, and the rest of the small bones of the nasal passages show signs of response to inflammation. These changes are the characteristic suite of '*facies leprosa*', and were described by Danish researcher Vilhelm Møller-Christensen from archaeological remains; as a thank you, the syndrome is named after him.* The damage to the nasal region is nothing compared to the damage a leper's sneeze can do – one estimate has 20,000 leprosy bacilli in a single blow of the nose.

The history of leprosy as an infectious disease goes back at least to the beginning of medical writing. The *Sushruta Samhita,* Indian medical texts from sometime in the first millennium BC,† mention a disease that may well be leprosy, as do texts from China; historical reports of what seems to be leprosy are reported in Greece not long after. Bioarchaeological evidence of early cases comes from Dakhleh Oasis in modern-day Egypt during the Bronze Age. Archaeologically, it seems that leprosy was yet another gift brought to the Americas by Europeans; there are no observed New World cases with the characteristic changes described by Møller-Christensen prior to the fifteenth-century contact period.

A great deal of what we know about the historical experience of leprosy comes from its prevalence in medieval Europe, particularly in northern Europe, where the number of cases and archaeological finds coincide to give the most comprehensive perspective on leprosy in the past. We will talk here specifically about the European case, because leprosy in medieval Europe becomes no leprosy‡ in post-medieval Europe, and that is something to think about.

Before the reader decides never to leave the house, there are a few common misconceptions that I should clear up. It's actually very difficult to get leprosy. It takes fairly constant exposure to the bacteria in the form of exhalation

---

* And you thought doctors competing to have fractures named after them was bad.

† They also provide earliest mentions of other diseases, *e.g.* scurvy and smallpox. Unfortunately, each ancient symptom has been attributed to just about every modern condition, so it's difficult to know which disease exactly any of the texts are referring to.

‡ Well, very limited.

droplets; you can't even get it from skin-to-skin contact. Exposure also doesn't guarantee infection – it helps to be malnourished or otherwise immunocompromised. Even then, you might get the tuberculoid form of the disease, which, while not exactly nice, restricts its actions to soft tissue. If you do manage to pick up a full-blown lepromatous leprosy infection, congratulations! But you are still more than likely going to die of something else first. Even if the disease does take hold, it can take years or even decades to experience the full swath of skeletal deformities, so archaeologists finding any lepers at all, is quite frankly miraculous.

Thank god then for leprosaria. Most of what we know about the skeletal effects of leprosy in the past comes from the concentrated remains from medieval hospitals that, among other functions, served as refuges for lepers; they also provided location for burial for those so afflicted. Despite the limited contagion of the disease, it was sufficiently horrifying that sequestration of those afflicted seems to have been a key feature of the response to leprosy. While isolated cases can be found filed among the wider population, the pronounced skeletal changes associated with leprosy are mostly identified in archaeological contexts of leprosaria. The nomenclature for these institutions comes from the same maudlin humour that pronounced the leper dead to the world, associating the afflicted with the biblical Lazarus. Maudlin is, as any Oxbridge scholar can tell you, a form of 'Magdalene',* and therefore relating to the Saint Mary of the same designation; the association with leprosaria of Maudlin Hospitals is also due to the biblical story. Leprosaria, Lazar Houses or lazarettos functioned in medieval Europe to contain the frightful contagion of leprosy, and we can actually see quite a lot of the medieval sense of disease in the foundation, funding and final abandonment of these institutions. The

* I spent ages mispronouncing the name of the Oxford college Magdalen, pronounced 'maudlin'. You'll excuse me for finding archaic English ridiculous.

perception of leprosy as a punishment by God is easy enough to comprehend, but it seems to have taken on the connotation of sexual sin to some extent as well – perhaps another reason that so many leper hospitals were named for the Magdalen.* With lepers being perhaps the most visibly wretched of God's creatures in the medieval world, it's no wonder that they attracted considerable charitable donation. Founding a leprosarium did great things for your spiritual sanctity.

Between about AD 1100 and 1350, 300 leprosaria were founded in England. This seems to be the pinnacle of a much longer run of endemic *Mycobacterium leprae* infection on the island, which ran concurrent with similar epidemics in Scandinavia and Denmark. Throughout this period leprosy was a largely rural disease, as Charlotte Roberts and Keith Manchester point out in their 1989 synthesis, with close-living family members likely to spread the disease among themselves so that it might cluster in a family or village, but it never seems to have exploded to epidemic proportions as soon as it hit the city gates. Leper hospitals, as charitable institutions that must garner support, were located more conveniently near major population centres, but the destitute they took in were not necessarily urban dwellers. Historians are also increasingly finding that society's terror of leprosy was not quite what we imagine it to be. Carole Rawcliffe, who wrote the actual book on medieval leprosy in Britain, pointed out in a lecture at Gresham College in 2012 that the unmarried, healthy female inmates of one of London's leper hospitals were sworn in as religious sisters, though

> *... they were not very obedient and they were always being told off for being a bit lippy, and there were evidently quite a few instances of what the shocked inspectors termed 'carnal copulation'. Since the canons were given to fraternising with the sisters over a drink, this is perhaps hardly surprising.*

---

* There seems to be some elision of Marys going on in Christian theology; the sister of Lazurus and the former prostitute aren't necessarily the same Mary.

This does not sound like the canons of the Church, men of great spiritual and worldly knowledge, were overly concerned about leprosy.

If leprosy is mostly a rural disease, what place does it have in a book about cities? The answer lies in those hundreds of hospitals. Three hundred leper hospitals is a considerable number for such a small island, and it suggests that both leprosy and the chance to demonstrate Christian charity were very popular. There is no way of calculating how many of the inmates of these hospitals had clinical leprosy, as the historic practice of differential diagnoses could be more a matter of religious sensibility than science. But it seems fairly clear that people knew when you *did* have lepromatous leprosy, or we wouldn't find such concentrations of the skeletal evidence of the disease in the cemeteries of lazarettos. What we have to question is why, after the Black Death upended the lives of urban and rural dwellers (leper and non-leper alike) in 1348, were there no more leprosaria? What happened to leprosy?

One argument says that leprosy might not always have meant leprosy – while some inmates demonstrably had *Mycobacterium leprae* infections, others may not have. Many have argued that any sort of skin disease might count, and we certainly know that skin diseases were of major concern to the people of Britain: from Edward the Confessor right up to James II, monarchs were obliged to cure the ills of their people through the direct laying on of hands. This 'royal touch' was of course a direct sign of God acting through the monarch, and the tradition persevered for at least 600 years in England and France. Cleverly, these monarchical manhandlings were judged most effective against a disease known as the King's Evil, which many modern researchers have identified as scrofula: another form of *Mycobacterium* infection that affects the lymph nodes, particularly around the neck, so it might be readily visible, but rather handily is quite likely to go away on its own. Queen Anne is the last British royal to indulge the custom, laying hands on the famous writer Samuel Johnston to cure him of his scrofula on

30 March 1712.[*] The death of the tradition, which seems to have attenuated steadily from the fifteenth century until finally coming to an end in the seventeenth, follows the trajectory of decline in the leprosaria, but for two things. First, the seventeenth century saw a clamour for *more* royal touching, not less; and second, the *Mycobacterium* that causes scrofula is not *Mycobacterium leprae* but *Mycobacterium tuberculosis*.

Tuberculosis is a disease that many people associate largely with wan Romantic poets and Victorian slums; this would be a mistake. In 2014 the World Health Organization estimated that there were almost 10 *million* new cases of tuberculosis, and over 1 million deaths. Tuberculosis, like leprosy, is still very much at large in the world, and what's more, we're running out of ways to cure it. Drug-resistant TB has been a fact of life for nearly 20 years, spurred on by the resurgence of tuberculosis infections in high-risk populations (the immunocompromised, intravenous drug users), and it's getting worse. Drug-resistant TB has arisen partly through our cavalier attitude towards taking complete courses of medication,[†] but probably most significantly through our even less excusable cavalier attitude towards providing full treatment for those most at risk. Extreme multi-drug resistant TB (XDR TB) is one of the first signs modern medicine has had that the golden age of antibiotics may not last forever; the Centers for Disease Control and Prevention in America give odds that XDR TB can be cured as low as 30 per cent. Still, the numbers of cases that affect the general population are low, and lack the tragic romance of the last great flourish of tuberculosis infections in the

[*] Either it didn't work, or it didn't work fast enough – Johnson had a subsequent operation and remained scarred for life.

[†] Stopping a course of antibiotics 'because I feel better now' is akin to handing infectious disease a loaded gun; you just might not realise it because the gun is pointed at a malnourished child in a country you can't pronounce. Don't.

eighteenth and nineteenth centuries, because they affect children in Africa and not Emily Brontë.

Tuberculosis, or the disease that we think of as tuberculosis, is caused by infection with *Mycobacterium tuberculosis* or, in very rare cases among people who spend a lot of time with certain animals, *Mycobacterium bovis*. Droplets containing the bacteria exhaled into the air spread the infection, and once inhaled, bacteria move into the lungs of a new host. In healthy individuals, they can even stay there, essentially walled off from the rest of the body by a high-functioning immune system. Given a chance, however, the bacteria can spread in three different ways: into the central nervous system, causing tubercular meningitis; into the lymph system, causing scrofula; or into the bone. It's this last path that provides the skeletal evidence of the disease we can identify in the archaeological record. As the TB infection spreads through the body it prefers certain types of bones, though any port will do in a storm; however, there are a few characteristic lesions that can be used to narrow down identification in archaeological skeletons.

Tuberculosis is a disease that takes away bone, rather than putting bone down. Combined with the characteristic locations of tubercular changes in the skeleton, this makes it one of only a few disease conditions bioarchaeologists can identify in the past. One key clue is angular kyphosis, or the angulation of the spine, which is the technical term for what happens when the little round stacks of vertebrae that keep you upright change shape. The erosive lesions of tuberculosis infection can break down the bodies of the individual vertebrae so that, instead of Camembert roundels, they look like pointed wedges of Brie. This pitches the remainder of the spine forwards, creating a distinct angulation. This characteristic deformity is identified as 'Pott's disease', after the surgeon who described it in 1779. Tuberculosis can leave faint traces on the bones of the chest, particularly the inner surfaces of the ribs and sternum – anywhere where there is nice marrow for the bacteria to proliferate. Just as in the metabolic diseases discussed in Chapter 5, the proportion of marrow within bone changes as the body grows, so that children frequently show evidence of tubercular changes to the hands and feet that are far less common in adults. Infection can also cause abscess formation, and these degenerative lesions force whatever bones are nearby to respond by forming erosive divots. Degeneration of the joints, particularly the hip, can also be observed; however, it's not always possible to say whether an eaten-away hip joint and associated smashed-up femur are the result of tuberculosis or some other sort of disease.

The archaeological evidence for the antiquity of tuberculosis currently extends far back into the human past, though perhaps not quite as far as some have claimed: the frequently cited case of a 500,000-year-old *Homo erectus* adolescent does not seem to hold up to the standard of the pattern of bony changes that we have discussed above. There are more securely identified changes in skeletons from the Neolithic period onwards in Europe and the Middle East, as well as historical accounts of the disease that are fairly unmistakable from the Bronze Age onwards in Asia and Europe. Both classical Indian and Chinese medical texts mention tuberculosis and its symptoms in relatively unambiguous form; this extends back to the *Huang Ti Nei Ching* texts from China, which have a semi-mythical origin date in the third millennium BC, but are likely younger. The skeletal evidence of tuberculosis in Asia is rather more frustrating. It has been identified at least as far back as the collapse of the Harappan civilisation in the second millennium BC. Many of the early identifications of tuberculosis came from forensic investigations of archaeological soft tissue, which of course comes largely from one source: mummies.

In Egypt there is a long tradition of encountering tuberculosis; in 1825 the first 'scientific' autopsy of a mummy was carried out on Irtyersenu, a woman who was buried in the cemetery at Thebes sometime around 600 BC. The colonial consumers of Egypt's mummy export empire in the eighteenth and nineteenth centuries were very keen on this sort of forensic investigation; mummy 'unwrappings' were cultural events; you could buy your ticket and watch a celebrity Egyptologist denude a 5,000-year-old body.* For fun. Almost two centuries later genetic tests identified *Mycobacterium tuberculosis*, which explained the pockets of fat found in her body as indicative of wasting rather than the assumption of corpulence by the original team.

The difficulty with identifying tuberculosis is that not all cases develop the skeletal lesions that bioarchaeologists look for; additionally, idiosyncratic immune responses might lead to the development of more or less unidentifiable cases. Suspected cases of tuberculosis have been reported from a variety of archaeological contexts, rendered all the more contentious by the ongoing argument between researchers who saw TB as an 'Old World' disease that was brought to the New World by Europeans in the fifteenth and sixteenth centuries, like leprosy. We can now decisively put this theory to rest: DNA has confirmed the presence of tuberculosis in 1,000-year-old bones from Peru in the mid-1990s, and further evidence has continued to roll in since. In the Old World, the first known case of tuberculosis is identified right at the cusp of the Neolithic transition at the cave site of Atlit-Yam in modern-day Israel. The 9,000-year-old bones of a woman and a year-old child were found with skeletal lesions that suggested TB infection; the child had reactive expansion of finger bones and lesions on the skull, while the evidence from the woman was not diagnostic. Molecular archaeology, however, stepped in to confirm that both skeletons contained *Mycobacterium tuberculosis*.

As in so much infectious disease research, genetic analysis of the pathogen offers an increasingly nuanced vision of the antiquity of disease in humans. Recent estimates suggest that *Mycobacterium tuberculosis* came out of Africa with modern humans some 40,000 years ago. After 10,000 years or so it split into two clades, with one group affecting only humans

* Either academic lectures were a lot more fun in the past or life before binge-watching was intolerably dull.

and the other group affecting humans and other mammals. Unlike the diseases that we have blamed on cows in the past, in the case of tuberculosis it's all on us: we appear to have given rise to the strains of *Mycobacterium* that affect mammals. The estimations of genetic timing provide a fascinating insight into our co-evolution with a killer. The strains of human-affecting tuberculosis that affect East Africa, India and Asia last had a common ancestor about 14,000 years ago – the approximate kick-off time of the Neolithic Revolution on those continents. The specific Beijing strain of modern tuberculosis infection has its origins in a genetic split from other branches around 6,000 years ago, during the Neolithic period of little farming villages around the Yangtze River. The Latin American strains in turn last split from their common ancestor around 7,000 years ago, suggesting that there is no reason to doubt the presence of tuberculosis in the New World prior to European contact. In fact, *Mycobacterium* DNA that seems to hover closer to the older branches of the lineage, *Mycobacterium africanum* and *Mycobacterium tuberculosis*, has been confirmed in a 17,000-year-old bison from North America, leading those researchers to suggest it as a potential origin for human TB infection rather than the other way round. However, the majority of the evidence emerging from the genetic profiling of the *Mycobacterium* complex of bacteria is that the human-transmitted types are older than the ones that we can pick up from animals; *Mycobacterium bovis*, the tuberculosis we unleashed on unsuspecting cows, seems to branch off around the time of the really intense domestication of the species around 5,000 years ago. Once you start looking at the molecular clocks, it's very easy to see our fingerprints all over this disease.

Having established the antiquity of tuberculosis, it's possible to trace it through history to the point where it becomes endemic, and then, suddenly, violently epidemic. In the burgeoning urban centres of the world, a triad of preconditions for epidemic infectious disease were being perfected: (1) gross poverty and associated malnutrition, which lowers immune competency and makes for a large pool of potential florid expression of infectious disease; (2) dense populations, which percolate and spread infection; and finally (3) connections to the wider world. The Roman cities of the very first centuries AD seem to have coincided with an uptick in the evidence of skeletal identification of tuberculosis – perhaps not an unexpected finding given the well-networked and urban nature of the Roman Empire. Interestingly, however the majority of finds are from the rather peripheral island of Britain. It's not clear whether this suggests a preponderance of tuberculosis, perhaps relating to the potential for the disease to expand into a naive British population, or the propensity of the British to excavate Roman skeletons and search them for signs of TB: most probably the former, as TB has been found in Iron Age skeletons from the UK as well as from a variety of sites just on the edges of the roman World, both geographically and chronically. The age of empires is a good one for tuberculosis: the great urban centres at their hearts make excellent incubators for TB, and their vast networks of trade and exchange pump the disease into new territories.

The network of urban cities in medieval Europe was a burgeoning disease incubator, as it found out to its great sorrow in 1348. Leprosy staggers off stage in the latter part of the medieval period, and is rare indeed by the early modern period of industrial and proto-industrial cities. Tuberculosis, on the other hand, while obviously present during the medieval period from archaeological and historic evidence, waits until the Enlightenment to really kick off. This seeming tag-team effort by *Mycobacterium* species to dominate the European field initially suggested to many researchers that what we see in the decline of leprosy and the rise of TB is in fact evidence of growing cross-immunity. It was noted very early on that many patients with leprosy actually die of TB; conversely, exposure to TB seemed to confer some sort of immunity towards leprosy. Could the spread of tuberculosis around Europe by the exponential growth in urban populations and the trade networks that connected them actually have inoculated the population against the scourge of leprosy?

Recent analysis of archaeological skeletons seems to say no. A team led by microbiologist Helen Donoghue sampled archaeological remains with characteristic skeletal changes attributed to either leprosy or tuberculosis infection from a range of sites spanning fourth-century AD Egypt to sixteenth-century AD Hungary. Their goal was to obtain genetic signatures from any *Mycobacterium* present to see if the cross-immunity theory could be supported. Instead, what they found was that, out of 32 samples, 10 had traces of both *Mycobacterium leprae* and *Mycobacterium tuberculosis*. In order for the team to find the bacteria from samples in the bone, the *Mycobacterium* infections would *both* have to be actively coursing around the body, leading the team to conclude that cross-immunity was unlikely. Instead, they argued, co-infection is a much more likely scenario, and one unfortunate consequence of having leprosy is a propensity to acquire and die of other infections.

The complicated interplay of tuberculosis infection and the decline of leprosy might then be more succinctly ascribed to the triumph of the high-mortality infection in a new ecological niche: the urban population. While tuberculosis was a lingering disease, with years required before the characteristic lesions appeared in the skeleton, it certainly didn't hang on as long as leprosy did, and its spread was far more rapid. It seems that at some point, perhaps around the fourteenth century, the transmission of infectious diseases ramped up to the point where virulence outbid its rivals in disease forms. Epidemics of infectious disease became *endemics*, naturalised citizens of cities thanks to the contributing factors of population growth, mobility and integrated networks for the transmission of people, goods and animals, all driven by the engines of urban growth.

It may be that how we think about disease in the past is not entirely unrelated to how we think about disease in the present. Rawcliffe suggests that much of the history of leprosy in the medieval period was written at a time when historians themselves occupied a world of plague and newfound pestilence. Empire had opened up the farthest-flung reaches of the world to the Europeans, and while there may not have been dragons, there were certainly diseases long thought dead. She points to the flurry of excitement over the scientific identification of the *Mycobacterium leprae* bacteria by Hansen in 1873 as evidence that historians were looking at the past through a pandemic lens. I would add that the 1870s was a time when the punishing regime of the dense urban megalopolis that London had grown into started to become apparent to those who governed it. In a city full of killer epidemics, it's easy to see how a terror from the past might be magnified. Whether leprosy was an intolerable burden of isolation, madness, disfigurement and disability, or whether this burden could be borne with the help of some ecumenical wine and company, we may not ever entirely know.

What we do know is that, in a newly densely populated urban world, the course of our health was forever altered. We have changed the environmental reservoir for the diseases that have trailed us for millennia, and we continue to do so today. Part of that is a function of the increased connection between our population centres, the roads, the sea lanes, the paths forged into new and ever more exotic disease habitats. Ebola thrives because it comes on four wheels and not two shaking legs; coronavirus spreads globally because the Hajj can only be made to one place but the faithful must come from any continent they find themselves on. The metal ages of bronze, copper and iron undeniably had their long-distance social and economic connections, often very impressive ones, but it's in a globally connected world that we see the true consequences of dense urban living.

Humans have undergone a series of 'epidemiological transitions', where the nature and type of diseases most prevalent in our species shift. The most commonly discussed is what many would argue is the first 'true' transition, occurring in the Neolithic when the concentration of humans and animals in settled locations allowed for the transmission of infectious diseases that would otherwise sputter and die. Tuberculosis is perhaps the clearest example of a disease that we do not see molecular or archaeological evidence of before the Neolithic, but bears all the hallmarks of a density-dependent disease. After appearing on the scene 9,000 years ago, it skulks through the bïoarchaeological record before once again making a nuisance of itself in the urbanised world of imperial Rome. Endemic if not epidemic, it knocks around for centuries before the huge reorganisation of human populations into increasingly connected, densely populated cities creates the circumstances for crisis mortality – plague.

CHAPTER TEN

Bring Out Your Dead

Like many bioarchaeologists, I have a fondness for plagues. They upend the natural order of things, cutting across the normal risk factors for ending up in archaeological samples and giving a snapshot, captured in death, of not just the old and the infirm but also a sample of the whole (unlucky) population. The tragedy of mass causalities exposes lives that would, statistically, rarely be unearthed, including the adolescents and adults who form the bulk of a living population, so rarely represented in a cemetery. Calamities such as plague that knock everyone into the grave with one indiscriminate sweep are one of the few chances bioarchaeologists have to overcome something known as the Osteological Paradox, a term coined by researcher James Wood and colleagues to cover the very awkward point that, in studying past lives, the evidence bioarchaeologists actually have to go on are past *deaths*. Without access to modern medical care, the greatest potential for mortality comes in old age and in infancy and early childhood. Death is less of a risk for adolescents and reproductive-age adults, until something comes along to level those odds.

A giant, sweeping epidemic disease that carries away huge swaths of population, respecting neither rank nor righteousness, the phenomenon of plague holds a particular cultural fascination. The death's heads and dancing skeletons of fourteenth-century European art reflect the macabre upending of order in the European world in the wake of the Black Death, with death itself brought front and centre in the narrative of European civilisation. Indeed, the after-effects of the incredible loss of life seen during the Black Death had far-reaching consequences, whose scope and scale researchers are only now beginning to unravel.[*] The Black Death didn't just fell kings; it felled kingship and the entire medieval way of life in the parts of Europe that it devastated. Increasingly, scholars are seeing the influence of plague in other areas as well – lesser-known consequences of the catastrophic mortality that rippled out from epicentres in Asia and Africa. But why should any one particular disease have such a virulent effect? What circumstances conspired to kill uncounted millions on three continents? What, exactly, is plague, and how did we get it?

Plague is not one disease, even though most people's first association might be the bubonic plague caused by *Yersinia pestis*, popularly known as the Black Death. Plague can be any disease[†] that strikes enough people with sufficient virulence to kill in large numbers. Your definition of 'large' might understandably vary depending on whether those numbers are made up of your friends and family, but it's another one of those things where you would know it if you saw it. Epidemic diseases require large numbers in order to succeed in the way the Black Death did. It does the bacteria or virus no good if the host organism keels over all alone in an isolated field without transferring the infection first. Some infections have a sufficiently long incubation period, or can hang about for long enough in nearby animals, that they can bridge the gaps between little clusters of people, forming a chain of infection that can rapidly spread across the landscape. The two requisite conditions for a local epidemic disease to become a pandemic are then very much the conditions bred in cities: dense populations, and well-worn roads between them. What we still don't perfectly understand, however, is why some infections make that jump – infectious disease is nothing new, so why does it sometimes take us so much by surprise?

---

[*] This is an academic translation of 'argue violently about'.

[†] Or, in some slightly less credible accounts, rains of vertically dislocated amphibians, grasshoppers, *etc.*

The real contribution of the invention of the city to the epidemiological transition is the shift from infection to *epidemic.* Infections are a nuisance, maybe even fatal; epidemics will kill you and leave no one left to bury the body. You cannot have an epidemic without a large, vulnerable population to infect: that's what makes it an urban disease. It's not that the diseases that cause epidemics don't strike outside cities – most of our modern epidemic diseases are actually rural infections direct from the rustic goodness of the soil. They would stay there, too, but in a networked world of roads, buses, cars and people, they now reach the dense urban populations they need to explode. It's the ready transmission of infection that makes a disease into a plague; the networks of roads, trade and the people who link an increasingly urban world. It's the susceptibilities of unequal city lives, however, that allow plagues to flourish.

It's possible to break the infectious diseases that kill humans down into different categories; these might be dependent on transmission vector (what you get it from), disease agent (bacteria or virus) or even climate and geography (it's difficult to get malaria if malarial mosquitoes aren't around). I'm going to cheat slightly and divide the major infectious diseases by the level of terror they inspire – that and how they are identified in the archaeological record. This means that some of the most insidious killers, diseases caused by contaminated food and water like cholera, will be discussed in Chapter 13, when we talk about endemic diseases in the context of Industrial Revolution. Syphilis is sufficiently macabre and mysterious that it gets its own chapter. In the previous chapter, we covered the peculiar case of tuberculosis and leprosy as a function of population density; much of the same reasoning will be used in the discussion here: people (lots of people) offer a great opportunity for a disease on the make to become a full-fledged plague.

What makes a plague? Dead people – and lots of them – in unusual places. As we go wandering back in time, there are few candidates to fit this description: the disease we know as Plague-with-a-capital-P, which we will get to, and the plague that came before it, smallpox. Smallpox manages to be both the oldest archaeologically known epidemic and by far the deadliest recent epidemic. It is caused by infection with one of two species of virus, *Variola major* or *Variola minor.* The latter causes alastrim, a relatively benign form of dermatological disease which is rarely fatal and is endemic to parts of Central America. The former has killed by some estimates 500 million people, and leaves a disfiguring rash in many of those who were lucky enough to survive; this is the disease we are interested in. Smallpox is a disease that acts first on the skin: there are spots, then there are fluid-filled blisters. These blisters are what cause the horrific scarring associated with smallpox.* Subsequently, the infection can progress even into bones, causing osteomyelitis, though this occurs only in a very small percentage of cases.†

Osteomyelitis is a very broad category of pathology; the term is used to describe any infection of the bone itself. It is the one sign accessible to bioarchaeologists who work on bone of potential smallpox infections. When bone is infected, it has an inflammatory response just like other tissues of the body, met in turn by a repair response that results in characteristic changes to bone shape. The infection may simply be evident as reactive bony changes on the interior or exterior surfaces of the bone. It can form an abscess, a pus-filled lesion, which may force the formation of a channel all the way out through the bone, along which said pus can suppurate. The clinical term for this channel is 'cloaca', which is Latin for 'sewer' and every bit as disgusting as you might imagine a Roman sewer to be. New bone formation on the surface in response to the infection can change the width or thickness of the bone, so when compared to an opposite number, say from the other side of the body, the shape is very much changed.

Of course, the majority of cases of osteomyelitis are not caused by smallpox but by a host of bloodstream-borne

---

* I do not recommend image-searching this.

† Though variable, 2 to around 35 per cent.

infections, with about 90 per cent of modern infections caused by the bacterium *Staphylococcus aureus*.

Clinical reports from the twentieth century give us insight into the varied courses smallpox itself could take; in an epidemic involving some 1,400 cases in what is now Malawi, primarily children were affected, first by the typical skin rash but secondarily by bone infections. The infection rates varied with the ages of the children, but seemed to have no link to any concomitant conditions (malnutrition or red blood cell diseases like sickle-cell anaemia); two-thirds of the children between the ages of one and five showed that the disease affected their bones. This may be similar to the pattern of skeletal involvement in the metabolic diseases discussed in Chapter 5, where the effect changes depending on which parts of the child's skeleton are most actively growing. In smallpox, modern clinical practice reports changes occurring in the elbows, the fingers and toes, and the ankles. The only known potential archaeological example of *Osteomyelitis variola* (smallpox skeletal damage) was reported by Mary Jackes from a seventeenth-century cemetery in Ontario, Canada from the bones of an older man who died around the 1650s. She traces the potential route of infection through a series of local indigenous interactions, encounters with colonial powers and all the way back to a London outbreak of 1628. It's not clear, however, that the osteomyelitis described necessarily results from smallpox; the lesions on the man's skeleton could well have resulted from an infection of bone picked up from other sources.

When did we get smallpox? It's a very interesting question. Historical descriptions of a disease that seem very much like smallpox come from India in the first millennium BC; medical texts attributed to the work of the rather mythical personage Dhanvantari record a disease of fevers, pustules and skin 'that seems studded with rice'. A further text attributed to Dhanvantari describes a process of inoculation that would not be discovered again until the middle of the second millennium AD. Reports of an inoculation – essentially, introducing the infection deliberately in a different or weaker form – process of similar antiquity in China have been suggested to be seventeenth-century exaggerations,* but the disease seems to have been attested in text by around the same date.

The key to preventing smallpox, and the clue to its eventual eradication, was the practice of 'variolation'. Variolation involves inoculating small children by scratching them with material taken from the blisters of cowpox, a bovine-focused relative of human smallpox, and seems to have been widely attested in Africa, Europe, India and China from at least the late medieval period, though likely practiced well before. More surprisingly, given the state disease theory was in at the time, it largely worked. A man called Edward Jenner came across the idea at the tail end of the eighteenth century from chance† observations of milkmaids; his experimentation led to the Western medical establishment's embrace of the inoculation concept, and eventually to the smallpox vaccine.‡

Historical records are all well and good, but they suffer from a number of inadequacies. First, for the vast majority of human history no one wrote anything down. Second, diagnoses of medical traditions which rely heavily on religion, spiritual beliefs and fairly out-there theories of the human body may not be easily comprehensible to the modern researcher.§ Despite smallpox only rarely leaving telltale signs of bony infection, bioarchaeology has another avenue for detecting the disease in antiquity: mummies. In 1979 physician Donald Hopkins was granted special permission to examine the upper half of the

---

* Fabrications.

† Apparently, also prolonged and deeply contemplative.

‡ 'Vaccine' was originally the term for only the smallpox vaccination; it derives from the Latin for 'cow'.

§ This even applies to diagnoses less than 500 years old; it took a chance encounter with deer-hunting terminology for me to realise that the frequently recorded early modern British cause of death termed 'rising of the lights' was a description of lung disease, and not some sort of beatific spiritual event.

unwrapped body of the Egyptian pharaoh Ramses V, who died in 1143 BC.* The mummy's skin showed evidence of a pustular rash, very akin to smallpox, and subsequently Ramses V became famous not for a perfectly sensible large-scale tax survey he carried out in his limited reign, but for being the first archaeological evidence of smallpox.

Due to the antiquity of historical records and the singular case of Ramses V, which was never confirmed in the lab, many historians and archaeologists have speculated that smallpox came out of India, or possibly North Africa, and from there circulated to China and to Europe. A great number of plagues recorded by history but undiscovered by archaeology have been attributed to smallpox: one among the Hittites of Anatolia in the mid-fourteenth century BC; the Plague of Athens in 430 BC recorded by Thucydides; the plague that struck the army of Alexander the Great in the Indus Valley in 327 BC; the Antonine Plague that decimated Rome and its provinces in AD 165; and, the Plague of Cyprian, a hundred years later. Not all of these are in fact confirmed smallpox epidemics: the Hittite plague has been argued to have been *Francisella tularensis,* and the Athenian one typhoid. Mentions come in from Syria and France in the first half of the first millennium AD, and after that smallpox is more or less an established disease everywhere but the Americas and Australia. It was established primarily as an *endemic* risk and only becoming *epidemic* on the occasions it encountered a new, immunologically naive host.

This is precisely what happened when *Variola major* swept into the Americas on the back of European refugees, speculators and adventurers. In virgin territory, where the population had no inbuilt resistance to infection, smallpox became the deadliest epidemic this planet has ever seen. In the conquest of the Americas by European colonial powers, the mortality rate for the native population was anywhere

* He has more names, all of which translate to variations on the theme of 'honourable', 'rich', 'long-lived' and 'strong like a bull'.

between 30 and 70 per cent. Of course, immunity is relative; globally, even in regions that have had smallpox infections for thousands of years, some 300 million or more people were killed by smallpox in the twentieth century alone as populations shifted, changed and crowded together. One important question remains, however. If *Variola* viruses exist in a killer form that came out of Europe and not-so killer form that was already in the Americas, and you can cure the first with a related virus that doesn't even come from humans ... why did so many Native Americans die? Why did the precepts of inoculation not work between the *Variola major* of the Old World and the *Variola minor* viruses of the New?

Perhaps this is because *Variola minor* wasn't a 'New World' version of the disease, but rather another form of the same Old World disease. It's only recently, with aDNA, that we've been able to more clearly understand the interaction of infectious diseases in the past. Research into the genetics of infectious pathogens themselves can reveal an astonishing amount, including variation in pathogen genetics around the world. By calculating the molecular clocks of the change-over in that material, researchers can even estimate the point in time that different strains diverge from each other, potentially becoming more or less virulent, affecting new hosts or even being geographically isolated. The *Variola* virus can be opened up to reveal just this sort of information. The complicated history of smallpox is a story of multiple different strands appearing on one continent only to disappear and appear on another. The low-fatality version that causes alastrim in the Americas shares a recent common ancestor with the low-fatality version of the disease identified in West Africa, suggesting that early smallpox cases in Brazil and the Caribbean were due to contact through the slave trade. Reconstruction of the disease suggests that it switched over to human hosts at least 16,000 or so years ago, spread out of Africa and conquered Asia in epidemic form before turning back to the West and invading Europe and North Africa. Pathogen genetics are increasingly rewriting what we thought we knew about the history of disease and civilisation; this is nowhere more true than in the case of plagues.

Because this is a chapter on plague, the reader will rightfully expect to hear a bit about *the* plague, the disease that caused the Black Death. Well, *the* plague has two black marks against it in my book: first, everyone else has written about it, and some quite well; second, infection with *Yersinia pestis* does not leave appreciable traces on the skeleton, and so makes for annoying bioarchaeology. Luckily for you, the third factor of plague more than makes up for all of the above: it kills a *lot* of people. Quickly. Faster than even smallpox, the rapid deterioration that follows infection with plague means that the epic mortality rates experienced do leave a scar in the archaeological record in the form of mass burials. They leave an even greater scar on the history and progress of society, as written and as experienced.*

The modern perception of the experience of the Black Death owes more to a cinematic imagining of biblical proportions than to the actual reality of a recurrent condition that (sorry) plagued Europe for centuries. Quite frankly, Monty Python and his Flying Circus hasn't helped either – think of plague, and the first cultural reference you'll have if you're of a certain age and a certain language group is Eric Idle banging a triangle and shouting: 'Bring out your dead!' How true to life (or death) were the not-quite-dead corpses that John Cleese heaped on top of a plague cart? Actually, there is considerable evidence for the level of devastation portrayed in films. Much of this evidence comes from contemporary written accounts, so while it might be supposed that there was a certain degree of dramatisation, there is no argument that the sweeping plagues of the fourteenth century were lacking in drama. *The Decameron* by fourteenth-century

---

* They do not affect nursery rhymes; the popular misconception that 'Ring Around the Rosy' was about the Black Death only appeared around 1951, alongside other novel inventions like the nuclear family and the H-bomb. Apparently the atomic future inspired fairly apocalyptic interpretations of the past.

Italian author Giovanni Boccaccio is perhaps the best-known contemporary account of the experience of the Black Death,* and it mostly concerns hiding from sick people and taking potshots at the avarice and moral turpitude of the Catholic Church.† Boccaccio relates the impact of the plague on Florence with a practical, if not clinical, air:

> *Not such were they as in the East, where an issue of blood from the nose was a manifest sign of inevitable death; but in men and women alike it first betrayed itself by the emergence of certain tumors in the groin or the armpits, some of which grew as large as a common apple, others as an egg, some more, some less, which the common folk called gavoccioli. From the two said parts of the body this deadly gavocciolo soon began to propagate and spread itself in all directions indifferently; after which the form of the malady began to change, black spots or livid making their appearance in many cases on the arm or the thigh or elsewhere, now few and large, then minute and numerous. And as the gavocciolo had been and still were an infallible token of approaching death, such also were these spots on whomsoever they shewed themselves. Which maladies seemed set entirely at naught both the art of the physician and the virtue of physic; indeed, whether it was that the disorder was of a nature to defy such treatment, or that the physicians were at fault – besides the qualified there was now a multitude both of men and of women who practiced without having received the slightest tincture of medical science – and, being in ignorance of its source, failed to apply the proper remedies; in either case, not merely were those that covered few, but almost all within three days from the appearance of the said symptoms, sooner or later, died, and in most cases without any fever or other attendant malady.*
>
> *The Decameron*, Giovanni Boccaccio, 1353

* Also the most amusing; and, as the only book I took on a three-month field season at Çatalhöyük, rewardingly long.

† There is also a lot of sex, quipping and combinations thereof.

Boccaccio was truly astonished by the disease that descended on his Florentine commune; while his words are the ones that have passed down to modern readers, they would have been sentiments echoed by his contemporaries. The most shocking thing, one gathers, was not so much the piling up of the dead, though Boccaccio describes how this appeared:

> *Many died daily or nightly in the public streets; of many others, who died at home, the departure was hardly observed by their neighbors, until the stench of their putrefying bodies carried the tidings; and what with their corpses and the corpses of others who died on every hand the whole place was a sepulchre.*

The shocking thing was that seeing a dead man, or two or three, being carried on planks to the grave through the street with no funeral procession around him occasioned all the emotion of, as Boccaccio himself put it, seeing a dead goat in the street.

The plague described by Boccaccio is the bubonic plague, or Black Death, that caused catastrophic mortality in Europe in the late 1340s. Many readers may be more familiar with the Black Death than any other subject of medieval history,* and can recite the details well enough: Plague came to Europe on a ship from India, it infected an enormous number of people by means of fleas on rats, some of those people died horribly with swellings (buboes) in their groins and armpits alongside a spotted rash, some of those people transported plague to other places, and in the end half of Europe was dead. For reference, at least the 'half of Europe' part appears to be correct, and the rest is only mildly misleading. Plague actually came to Europe several times, for instance as the Plague of Justinian in AD 541, but seems to have lacked the ability to transport itself quickly across the continent to infect

* Except for the large proportion of colleagues from archaeology and history who I expect will buy this book. At full price.

the whole of Europe.* No one is really sure how it got to Europe in 541, but it may have come through Egypt via trade with India over the Red Sea. For the Black Death of the fourteenth century, there are no recorded episodes of epidemic disease in India that fit. Instead, the origin was more likely to have been via the Silk Road from northern China, which seems to have had sporadic episodes of deadly epidemics from the 1330s. Admittedly, boats did seem to feature in at least a few outbreaks, as described below. Fleas and rats, meanwhile, get a bad rap; not all plague is spread by flea bites – the last outbreak of plague in China in 2009 was marmot-based. There are three pertinent varieties of *Yersinia pestis* infection: bubonic (buboes!), septicaemic (in the blood), and pneumonic (in the lungs). The latter is a mechanism of direct transmission from human to human, and no fleas need apply.

Plague was then endemic in Europe, as it probably had been in Asia, for at least the next 400 years, with occasional outbreaks. The diaries of famous man-about-historic-London Samuel Pepys record the experience of the London outbreak of 1665, where he notes that everyone has such a deep and abiding suspicion that Londoners might be carrying plague that he has to keep telling people he's from suburban Woolwich. For our diarist, however, plague did not seem too great a tragedy – it made him melancholic when confronted with actual corpses, but by New Year's Eve of 1666, Pepys is able to report that he has 'never lived so merrily (besides that I never got so much) as I have done this plague-time'.

It was in Pepys' time that one of the greatest leaps forward in public health, the publication of weekly mortality rates parish by parish, was instituted: the Bills of Mortality. Pepys, and his potential hosts, were able to keep up with the weekly stats on London plague deaths because of the widespread publication of those Bills of Mortality, the morbid factsheets of London deaths and their causes compiled by the government

*So it seems the collapse of the Roman Empire and the intransigence of the northern frontier were actually quite helpful, in that respect.

and read with no small frisson of sanctimonious *schadenfreude* by country and town folk alike. The Bills, and what can be made from their rather idiosyncratic data, are a fascinating subject that will be explored in greater depth in Chapter 13.

More archaeological wordage has been put into this subject than I could possibly recount, so it's enough to keep abreast of what we now know about the infectious disease caused by *Yersinia pestis*, a subject that has been radically transformed by the incontrovertible genetic identification of the bacteria from plague burials in the lab. In 2015 the team led by Simon Rasmussen and Morten Erik Allentoft of the Centre for GeoGenetics at the Natural History Museum of Denmark published a groundbreaking study of the genetic archaeology of the plague pathogen. *Yersinia pestis,* it seems, has been loitering around the steppes of Central Asia for some time. Animal and human versions have a long history in the region, with the human-infecting pathogen diverging thousands of years before the first ancient DNA evidence of infection. Their work showed that *Yersinia pestis* had been present for more than 5,000 years in Eurasia – but not as we know it. The Bronze Age version lacked a key genetic mutation that allows it to live in insects; it could not have been spread by fleas. The very oldest strains found may not even have been able to form buboes, but instead relied on pneumonic spread. The story is only becoming more complicated as further genomic work unveils a story of multiple origins and complex routes to infection and pandemic.

The very first allegations of biological warfare concern the deliberate spread of epidemic disease. The catastrophic epidemic that befell the Hittite Empire of Anatolia (in modern Turkey) around 1335 BC is tentatively attributed to an outbreak of *Francisella tularensis*. This is an infectious bacterium carried by animal vectors that can painfully and fatally affect humans. The texts of the period seem to suggest that whatever struck the Hittites went on to trouble everyone they fought. A very wandering and fanciful interpretation of the Hittite ritual release of a ram and heavily made-up woman undertaken before battle to appease the gods and stave off their hunger for human flesh has been interpreted as a mechanism (through either creature) for deliberate infection of enemy troops, which would make the Hittites the first-ever practitioners of biological warfare.*

One of the more colourful explanations for the origins of the catastrophic *Yersinia pestis* epidemic that ravaged the European and parts of the Asian continent in the fourteenth century AD starts with just such a story. Gabriele de' Mussi authored a memoir of the ravages of the Black Death in the years immediately after, in which he gives a supposedly† first-hand account of the siege of one of the great trading outposts of his native Genoa: the fortress at Caffa, modern-day Feodosia, in the Ukraine. The great Golden Horde of Mongol warriors led by Toqta Khan besieged the Italians at Caffa for some three years before eventually, according to de' Mussi, succumbing to a 'mysterious illness which brought sudden death', and which was at the time endemic to their territory. This disease caught up with the Mongols at the siege of Caffa in 1346. In an act of desperation, defiance – or just plain peevishness – the dying warriors began catapulting their dead over the walls:

> *... in the hope that the intolerable stench would kill everyone inside. What seemed like mountains of dead were thrown into the city, and the Christians could not hide or flee or escape from them, although they dumped as many bodies as they could into the sea. And soon the rotting corpses tainted the air and*

---

* If the ritual did occur as recorded, my sympathies are entirely with the woman and the ram.

† There seems to have been some confusion about de' Mussi's account, which his editor originally put forwards as an eyewitness testimony from the refugee ships that had spread the Plague back at home. It was later revealed that the medieval lawyer-notary had in fact spent the Plague at home in Italy. The amount of confusion owing to our best account of the European spread of the Plague being written by a lawyer, I leave to the reader's estimation.

poisoned the water supply, and the stench was so overwhelming that hardly one in several thousand was in a position to flee the remains of the Tartar army. Moreover one infected man could carry the poison to others, and infect people and places with the disease by look alone. No one knew, or could discover, a means of defense.

Gabriele de' Mussi, 1348

This seems to be a clear description of medieval biological warfare, designed not just to horrify but to kill. An even earlier example of the use of bodies to kill dates back to the tumultuous period preceding the reign of the Abbasid dynasty in the eighth or ninth century AD, in what is now Jordan. Archaeological excavations revealed the remains of at least six individuals who had been dumped in a then-full water cistern. Whether the intention was to hide the bodies or poison the city, the effect would have been the same. The very aspects of the city designed to protect its inhabitants – walls, water, stores – become engines of destruction.

The case of smallpox in the Americas could almost be excused as a terrible and naive accident, were it not clear that in a number of instances the colonists deliberately tried to spread the pandemic to the indigenous people. On 24 June 1763 the Pontiac tribe had besieged the colonial Fort Pitt (now Pittsburgh, United States), trapping a dense enough number of the European-origin population inside the fort that infectious diseases, like smallpox, began to spread. The Pontiac sued for peace on two attempts, with senior members of the tribe visiting their cornered enemies to try to convince the English to give it up already. We know this because, on that day, local merchant William Trent wrote an account of the visits, noting:

Out of our regard for them, we gave them two Blankets and an Handkerchief out of the Small Pox Hospital. I hope it will have the desired effect.

From the journal of William Trent, 24 June 1763

Trent was not alone in his hopes. The British army shared and indeed encouraged this; in a letter from about a year previously, the commander of the British forces in America, Sir Jeffrey Amherst, writes:

> *Could it not be contrived to Send the Small Pox among those Disaffected Tribes of Indians? We must, on this occasion, Use Every Stratagem in our power to Reduce them.*

Following this up about a week later with a commendation:

> *You will Do well to try to Innoculate the Indians by means of Blanketts as well as to try Every other method that can serve to Extirpate this Execrable Race.**

It is interesting to note in the devastation of smallpox that none of the quiet, contemplative morality of the earlier European plagues seems to be featured. Where the Black Death inspired extensive soul-searching, moralising and what can only be described as a crisis of faith across an entire continent, smallpox by the fifteenth century seems to have been an accepted fact of life, and one that could be turned to advantage by the enterprising imperialist.

Plagues, it seems, are endemic to our world. It's fascinating to think that plague – this terror, this awaited apocalypse that carried off such a huge number of mothers, fathers, brothers, sisters, children (and rats) – became conspicuously normal in medieval Europe. Much of the work on this chapter has been undertaken in the very pleasant German tourist resort of Überlingen, located on the northeast shore of the Bodensee (Lake Constance). I was supposed to be there recapturing family history and my sense of bewilderment at the German language, but as a sort of occupational hazard I spent quite a bit of time looking at the records of the town's rather significantly antique

---

* Humans, as I may have mentioned previously, can be terrible people.

hospital. During the mid-fifteenth to mid-sixteenth centuries of the Holy Roman Empire, Überlingen was a Catholic town with perhaps 3,000 inhabitants. Its *Seelhaus* was a 'soul house' for the indigent, particularly indigent women, and those struck by the 'evil pox' (syphilis). Überlingen is not particularly given to outbreaks of plague; indeed, its salubrious thermal baths have long been viewed as just the sort of environment to promote health. Nonetheless, even this small lakeside resort town had no less than three houses to accommodate the poor and the wretched, many of whom were, by virtue of the former characteristics, also the diseased. Überlingen, however, does not exist in a vacuum. Passing by the striking church of Birnau on a boat trip to the flowery tourist island of Mainau, my mother commented that it was there that her family went for John F. Kennedy's memorial service, making it a more worldly village than I had previously suspected. Out running, I passed a series of derelict train stations that had welcomed passengers from the industrial cities of the north to the thermal springs. Those train tracks follow the path of a Roman road that is millennia old. Tiny, adorable Überlingen has been trapped in the weft of the urban fabric of Europe for a very long time, and has the plague deaths to prove it.

My point here is that plagues are diseases of contact. Contact between different environmental reservoirs for diseases – and that's all we are to a bacterium – provides an opportunity for infectious disease to blossom into a full-scale pandemic in the absence of host immunity. To the *Variola* virus or the *Yersinia pestis* bacterium, our species are essentially little ambulatory nurseries – meals on wheels. The level of contact that preceded the transmission of the major plagues discussed here is very closely linked to the rise of cities and the push-pull effect they have on goods, the people who carry them and the armies that fight over them. The plagues of the second millennium BC arrive at the dawn of interconnected regions, when the Near East connects to Central Asia, the Eastern Mediterranean and possibly beyond. The age of big Eurasian empires (in the first millennium BC: think Persia, Rome, Mauryan India or Qin/Han China) that brings a wider circulation of goods, ships, people and attack elephants expands the range of diseases further still, and we see it in the accounts of the plagues of Athens, Rome and beyond. The devastation of the fifteenth and sixteenth centuries in the Americas is a clear extension of Old World plagues to New World bodies. And what we see in the modern day are the last bastions of infectious isolation blown away by globalisation, by the quest for intensive resource extraction in even the most obscure parts of the world. The microscope I use for my digital slides is made in Japan using American science, assembled in Germany and sold by Brits. This year alone I will go to Turkey, the US, France, Germany, Spain (twice),* Austria, Switzerland, Lichtenstein, Luxembourg, Belgium, Finland and Japan (twice). Our modern plagues are a product of our global network of cities and the states they represent. Like the plagues of the past, though, our modern plagues will probably fall into a pattern of cursory appearances and growing immunity, until they too are replaced by whatever new plagues await over the horizon when we find a way to expand our connected cities beyond their current range.†

Bioarchaeologically, we uncover evidence of plague from the genetic material locked in ancient bone; from collagen that remains trapped, usually in teeth, long after the body has skeletonised. Molecular archaeology is our primary weapon in understanding the pandemics that have shaped our population history, and in so doing shaped our political, social and economic history. A number of scholars have argued that the depopulation of Europe by the Black Death was an underlying reason for the uptick in the value of labour that finally freed many of Europe's peasants from their ties to the land. High labour value meant wages, which meant mobility, and mobility meant opportunity. The population of Europe did eventually recover from the fourteenth-century Plague, but it did so unevenly. The major cities drew in newly

* Missed Andorra by a literal mile. Disappointing.

† Michael J. Crichton may have been a visionary.

mobile workers while many rural enclaves shrivelled to nothing and disappeared. The disappearing towns of England and Denmark – DMVs or 'deserted medieval villages' – are well-known examples of how population movements away from rural life changed the character of the medieval world. Plagues may have spread through cities and because of cities, but in the end it is plague, with its attendant economic and social crises, that built cities into the truly monstrous sizes we see in the proto-industrial, early modern period.

CHAPTER ELEVEN

Tainted Love

One of the enduring curiosities of being a foreigner living abroad is learning the unspoken rules of your new homeland. It's all very well and good coming as an American to the United Kingdom, having the advantage of (mostly) understanding the language,* but there are all sorts of cultural minefields lurking unsuspected. One of the most inexplicable of these is the United Kingdom's relationship with the continent it sits adrift from, represented in almost all cultural references by the land you can *just* about see from the white cliffs of Dover. Over there, when you are over here, is France. Never mind that the territory has been held by interchangeable lineages for a thousand years; never mind the shared heritage, vocabulary … oh, no. Here, we are '*rosbifs*'; there, they are 'frogs'. Here, they serve French custard; there, they offer you *crème anglaise*. But what I find particularly amusing is the cultural insistence, on either side of the Channel, that one of humanity's most shameful, disfiguring diseases is obviously from Over There – the French speak of the *maladie anglaise*,† which the previous centuries of English speakers knew as the 'French pox', which the lab would call infection by *Treponema pallidum*, and which proper doctors would call syphilis.‡

---

* I'm sorry but that is *not* a biscuit.

† Hilariously, the eighteenth-century French theory of English diseases also holds that there are particular distempers common to the British Isles: consumption, scurvy, rickets and hypochondria. The first three would seem to excuse the fourth.

‡ If we'd listened to Oviedo, Spanish colonial adjutant and rather biased chronicler of the history of European expansion into the Americas, it would be called the 'American disease'.

The scourge of the last 500 years of sexual intercourse has of course been attributed to unsavoury neighbours by a host of countries: for Russians, it's the Polish disease; for North Africans, the Gallic; for Turks, the Christian; and a remarkable number of France's neighbours seem to agree that it's definitely the French disease. The name 'syphilis' is attributed to an allegorical poem involving an eponymous shepherd who pissed off Apollo; the disease is of course a visitation of divine retribution.* In a modern world ravaged by fatal sexually transmitted infections, it's all too easy to imagine the impact of syphilis in the popular imagination of the latter half of the most recent millennium. It's even easier to imagine given that syphilis flourished in a European world of texts, illustrations and satirical cartoons. Most of what has been written about the history of syphilis concerns its effects on that most self-analytical of continents, but of course the truth is that syphilis is a global disease. According to the World Health Organization approximately 79,000 people died of syphilis in the year 2012, more than half of whom died in Africa. While today's syphilis is a disease often encountered in conjunction with HIV/AIDS (another disease disproportionately affecting communities in Africa), the rest of the world need not feel so smug. With or without HIV infection, the American Centers for Disease Control and Prevention reports that syphilis rates in the United States have gone up 227 per cent in the last 15 years.

What is a chapter on a disease of ye olde soldiers and sailors doing in a book about cities? Aside from it being a fascinating bio-historical mystery, there is the inescapable fact that syphilis occurs in a globally connected world. Like the plagues of the previous chapter, the disease itself is not caused by city walls, social hierarchies or getting most of your calories from cereal crops. But the conditions that made syphilis such a sensation† are the result of urban organisation. Cities are the

---

* Some have suggested that the actual etymology derives from the Greek '*sus*' and '*philos*'. That'd be 'pig' 'lover'.

† Burning, itching …

beating hearts of trade networks, taking in, churning out and generally pumping people all over the place. Just as the Genoese welcomed the Black Death when their citizens fled home from the Silk Road fortress of Caffa, it was the driving pressure to discover new sources of the raw materials that power the engines of urban states that spread syphilis. Historical peculiarities like the Great Bullion Famine in Europe in the fourteenth and fifteenth centuries conspired to lead to the eventual importation of syphilis into Europe. In virgin* territory, it had devastating effects: not quite on par with the smallpox exported to the New World, but a virulent chancre on the … face of the continent none the less.

But is that the whole story? The history of syphilis has remained one of bioarchaeology's most contested topics.† If we want to unravel what syphilis means for the health of cities, we'll have to unpick the competing arguments for the antiquity (or lack thereof) of the disease. It may help, however, if we first give some idea of what syphilis actually is. Syphilis is the name given to the disease caused by infection with the bacterium *Treponema pallidum*. Treponemal diseases cause disfiguring or discolouring sores (or both) in the generally non-sexually transmitted diseases of endemic syphilis (or bejel) and the skin disease pinta. Bone is sometimes affected in the (mostly) non-sexually transmitted disease yaws, and if the infection progresses, very floridly affected in the two varieties of the disease that we know as syphilis: the sexually transmitted venereal disease, and the congenital variety passed on from mother to baby. Up until the last decade, it was thought that treponemal diseases were all caused by the same bacterium. Fascinatingly, genetic research has now unlocked one of syphilis's last dirty little

---

* Cough, cough.

† All academic disagreements are passive-aggressive and privately devastating. See for example the reviews of this book.

secrets: *Treponema pallidum* has four separate subspecies,[*] and each one has its own special disease to cause.

Starting with the infection that causes the least physical harm, pinta is a skin disease endemic to southern North America, Central America and the northern part of South America. Infection by the *Treponema pallidum* subspecies *Treponema pallidum carateum* causes a lesion to form on the skin, and as the disease progresses and more lesions form, the skin around them can change texture and finally colour. The skin of a later-stage sufferer can be almost completely mottled by the overlapping colour changes caused by these lesions; the conquistador Hernán Cortés may have been referring to the condition when he described some of the indigenous people he encountered as being of 'different colours, even in the same individual'. Pinta, like yaws, is a disease of the tropics, preferring warm, moist environments, and is spread by skin-to-skin contact – a useful trait in climates that are clothing-optional.

Endemic syphilis – non-venereal syphilis or bejel – is present in slightly drier climates, and caused by the subspecies *Treponema pallidum endemicum*. While it is more common in the Eastern Mediterranean and Middle East, along with West Africa, the disease has certainly occurred outside this limited region. The course of the disease is very similar to that described for 'proper' syphilis, bar the nature of transmission; non-venereal syphilis is usually transmitted orally through mouth-to-mouth contact or even sharing utensils, though any sort of contact between broken skin and a lesion will do. The initial lesion usually appears in the mouth, but the infection can spread to the skeleton and cause changes that are very difficult to distinguish from venereal syphilis. The bones affected by treponemal diseases are usually those with the least padding over them, so things like fingers and shins are commonly where the major modifications to the skeleton occur. The infection of the soft tissue spreads into the bones

[*] Five, if you count the one that only rabbits get. Rather unfairly, it seems that rabbits can get at least two of the human strains, while we cannot get the rabbit disease.

as osteomyelitis, the bone infection diligent readers will recall from the chapter on smallpox. The wonderful thing about treponemal diseases, however – from a bioarchaeological perspective – is that they cause a very specific type of problem: a lesion called a gumma. For the benefit of all, this can be summarised as an inflammatory response of bone to infection. The end result is that bone is eaten away by the gumma, though the response to inflammation can sometimes be observed in the deposition of new reactive bone on the outside surface of affected parts of the skeleton. These are the clues to identifying any of the treponemal diseases from the skeleton. The potential infectious changes to the bones in endemic syphilis are more or less identical to those described below for venereal syphilis, the only difference being that there seems to be a preference for affecting the tibia and ulna, and less so the bones of the skull.

Yaws, infection by *Treponema pallidum pertenue*, also causes pathological changes to skin, joints and bones. It is a tropical disease by nature, and seems to usually be contracted in childhood; this means that the age or state of healing of any changes in the skeleton may be helpful in diagnosing archaeological remains. The tibia is again a characteristic site of infection, and the amount of new bone deposited on the shin can make the bone look bowed, even though the underlying orientation of the bone hasn't changed. Once the infection has progressed to a sufficiently advanced stage, however, the skeletal lesions are essentially undistinguishable from those of venereally transmitted 'proper' syphilis.

Venereal syphilis is the star of our show here. This is the disease that makes its debut in the fifteenth century AD as a sexual plague of epidemic proportions, and runs rampant into the modern era. Infection by *Treponema pallidum pallidum* sets in motion a train of pathologies that stretch out over time; this extended process of the disease is what leads to the distinct bony changes that allow us to diagnose treponemal infections from the human skeleton. After all, if yaws, bejel or syphilis killed you as quickly as plague, your bones would have no time to react. The first stage of the infection is a localised lesion – a chancre; considering syphilis is venereally transmitted, I will leave it to the reader to imagine where these lesions might appear. The infection spreads to the bone as in yaws and bejel, and there it has a specific set of sites where it starts to act. The bone-eating gummas usually affect more than one site, and if one side of the body is affected, it's likely that changes will be seen on the bone on the opposite side as well.

The bone's reaction to the gummas is described as sclerotic, which can be conceived as a sort of bunching up of dense tissue around the edges of a lesion. This is what results in the diagnostic condition of caries sicca, where the bones of the cranium look like chewed gum that's hardened on a sidewalk. As with leprosy, the infection can mess with the bones of the nasal passages, and the nose and maxilla can be completely eroded away by infection. The long bones meanwhile can respond with new bone that forms very thickly over the surface of the old bone, with the appearance of worm-eaten wood. The tibia in particular seems to be affected; the curvature resulting from new bone has been described as 'sabre shin'. Eventually, the infection can move into the brain. Once neurosyphilis sets in, there are a host of reactions, psychological and physical, and loss of sensation can cause destructive arthritis in the joints from being used and abused without awareness of the damage being done. The end game, without treatment, is not a pretty picture: syphilitics could end up with holes in their bones and suppurating ulcers all over their anatomy. Perhaps eventual madness was something of a relief.

Congenital syphilis is the saddest of the lot, being a disease that affects the young, who have no say in the matter. It was originally thought that only the venereal form of the disease could be passed from mother to child, but there are suggestions that this can also happen in yaws and bejel during childbirth, if not through the placenta. The infection causes characteristic changes depending on the stage of growth of the foetus or infant; as discussed in relation to the difference between childhood rickets and adult vitamin D deficiency, the body is vulnerable in different ways at different ages. When it does not cause stillbirth, infants who are born with the infection have growing ends of long bones that are poorly formed and might have extra bits of cartilage that have calcified; these are not instantly diagnostic, however. Properly bowed shins, not ones that just *look* bowed, can also result. The characteristic (but, again, not universal) 'saddle nose' feature of congenital syphilis is caused by the collapse of the nasal bones under the weight of infection.

One characteristic – if not diagnostic – marker of infection is the changes in the front two baby teeth and the first baby molar. Through some mechanism, probably akin to the one that causes the lines on teeth discussed in the first few chapters (where tooth enamel growth stops for a bit before resuming), the very tips of the front teeth don't form, leaving a funny half-moon divot in the biting surface, called 'Hutchinson's incisor'. In the molar, the normal chewing surface bubbles from a handful of bumps into something resembling a mulberry; hence the name 'mulberry molars'.

In short, there are four-ish recognised types of treponemal diseases that affect humans. All but pinta affect the skeleton, and therefore leave evidence for the bioarchaeologist, but this is no guarantee of a correct diagnosis. The bony changes are so similar in yaws and the endemic, venereal and congenital forms of syphilis that a degree of thought has to be put into the identification of any particular subspecies of infection. Congenital syphilis and yaws infections are acquired in childhood, and so you might be able to restrict categorisation a bit by working backwards from evidence of healing and active infection to develop a disease chronology. Yaws likes to be warm and moist, while endemic syphilis likes to be warm and dry, so geography might give some clue, but this is less helpful once long-range travel becomes a possibility. Or, you might do as I suspect many well-meaning bioarchaeologists have done in the past, and just call anything with chewing-gum head and a tibia that looks like a sword 'syphilis'.

What is it about syphilis? Why is this syphilis still such a source of fascination for historians and archaeologists alike? It's a bit of a long story, and the argument has been going on for at least 300 years. 'In fourteen hundred and ninety-two, Columbus sailed the ocean blue.' This is the rhyme that all American schoolchildren were taught in my generation. Though the veneration of the man was well on its way out in my home state of California, the descendants of the colonial powers of the Americas had spent hundreds of years integrating the story of Christopher Columbus's voyage to the New World into the national identity; so we learned his name, his ships and his 'discoveries'.* While by the 1980s educators were allowed† to concede that rather a lot of people had suffered in the forming of the United States, this was generally put down to either an unfortunate and totally unforeseeable series of accidents, and/or the brutal march of progress. Noticeably absent from the discussion of the deprivations of this insane gamble by Atlantic powers to extend their reach to foreign markets were the stories of rape, murder and disease.

Of course, they never teach you anything interesting in school.‡ When Columbus arrived in Santo Domingo, he and his men came off their ships with their animals, entitlement and disease. The local Taíno people were immediately decimated by some mysterious infectious disease, and the process of European colonisation of the Americas had truly begun. Even Columbus's contemporaries had to go to quite some extent to justify their actions in expanding the power of their states into the Americas. Luckily for them, critical approaches to history had not yet been invented. If we want

* Columbus's most uncontroversial contribution may now be as a slang term for when Europeans discover something non-Europeans have known about for ages; it's called 'Columbusing'.

† But frequently chose not to.

‡ Just kidding, kids! Stay in school! A PhD is almost the only thing that can stave off the burdens of adult life, like jobs, money and being treated like a grown-up by your family.

something more than myths and manifest destiny, however, there are more practical (some might say factual) accounts of the Columbine Overtures.

History records the fate of quite a few of the men who felled the Caribbean with their imported diseases and acquisitive militarism, and one of them is of particular interest to our story. Martín Alonso Pinzón was the captain of, rather coincidentally, the *Pinta*. After the long journey back from the Caribbean, he made it as far as Spain before collapsing and dying, covered in sores. This was Patient Zero, or probably more accurately, Patient One-of-About-Fifty; it's not entirely clear how many of the seamen who had returned from the New World would have contracted similar diseases. Either way, Pinzón's case was the beginning of one of the most thoroughly documented diseases of the pre-modern world. Notes on his case were published in 1532 suggesting that the symptoms of the returnees of 1493 were clinically similar to the epidemic we would come to know as syphilis.

Pinzón was reportedly covered with boils by the time he collapsed in Galicia. Two years later and more than 1,600 kilometres (1,000 miles) away, the French siege of Naples went suddenly, horribly wrong.* The French soldiers rapidly developed ulcerating sores on their nether parts, followed by a rash of sores on the rest of them. This was widely attributed to a disease picked up from either local prostitutes or camp followers, depending on your level of French or Italian nationalism. What happened in the years between 1492 and about 1498 is the source of considerable debate. The more accepted modern consensus seems to be that the disease specifically identifiable as syphilis appears in 1493 in Spain with Pinzón, in Italy at the siege of Naples in 1495 and everywhere else soon thereafter. There is nothing in this timeline to contradict a theory of New World origins; call it *morbus gallicus, bubas* or pox, this school of thought holds that

* From an insider's perspective, it's very difficult to see how a siege could ever go right.

there are no convincing earlier historical accounts of the disease we specifically know as syphilis. Rival schools argue back that, *au contraire*, the early first millennium BC medical texts of China, Greece and India that we keep coming back to in these chapters describe things that could be syphilis.

So, while the majority of historical reconstructions of the event have the disease arriving in Italy in 1495 after spreading from the crew of Columbus's expedition to the Caribbean, there are alternative theories. A school of thought very popular in the mid-nineteenth century held that syphilis had been present in Italy since at least 1493, which does further complicate the timeline of its transmission if it's supposed to have got off a boat in Spain in the same year. If the various reports of syphilis in 1493 from parts of Germany and northern Europe were also valid, it would, as medical historian John Le Conte put it in his long manuscript on the subject for the *New York Journal of Medicine* in 1843, require 'an extent of licentiousness and a refinement of depravity, beyond the utmost bounds of credibility'. The alternate explanation for the origin lies in the expulsion of the Moors from Spain resulting in the movement of disease both to North Africa and to the southern shores of Europe,* where pockets of non-Christian refugees lived in sometimes wretched condition. Syphilis in this scenario was then transferred *out* to the Caribbean several years after the European epidemic of 1498, on Columbus's third voyage.

So, what if the disease had already been established in Europe? If the condition existed already in the Old World, why did it only reach such terrifyingly epic proportions in the 1490s? If a disease that started with genital lesions and moved on to full-body sores followed by madness and death had existed for a considerable time, you'd think it would at least have had a name. Those who favour the idea of an Old World origin for syphilis have an answer for these questions,

---

* Perhaps prompted by not-inconsiderable racism on the part of its most vocal proponents in the earlier part of the nineteenth century.

rooted in the theory of disease adaptation. Much like tuberculosis, syphilis is a disease that needs a certain density, level of immune system health and set of behaviours in its host population to flourish. Perhaps changing conditions at the tail end of the fifteenth century prompted a new and more virulent life cycle for an older venereal disease. The medical historian Ivan Bloch wrote in 1901 that syphilis had to have been imported from the New World, because the licentiousness of earlier ages would have spread the disease like wildfire if only it had been present in Europe.

There is an interesting counter-argument to this of course: that it was exactly the high-minded chastity and moral revolution of the late medieval period that caused such misery from a venereal disease. We know today that, of the limited number of people who are functionally immune to infection by the sexually transmitted HIV virus, some are working prostitutes along the main roads of Africa's transport hubs where HIV is most prevalent. Did shifts in disease exposure cause syphilis to really take off - perhaps sex had fallen out of fashion by the Renaissance? Alongside a revolution of private space that saw travel and accommodation go from the provision of public rooms and spaces to a trend for private chambers, this line of thought suggests syphilis only emerged when it did because the behaviours which had previously held it in check were no longer permissible – namely quite a bit of wanton fornication, something Bloch avers was happening *constantly* in the Middle Ages. Nonetheless, I think these speculations can probably be dismissed on the grounds that people of all times and ages have had sex, and reasonably frequently; witness the profusion of humans on the planet.*

It's incredibly difficult to establish with any certainty which of these conditions that we see in the historical record were in fact syphilis and which might have been any other form of disease. The clear difference between syphilis and

* I also don't think I'd call the time of the Borgias and the Medicis 'abstinent'.

other virulent poxes like smallpox is the venereal spread of the disease and its progression from a limited lesion to a whole-body illness. Plagues of people dropping dead covered in boils sound less like a suddenly virulent syphilis and more like smallpox; it's not entirely clear how many people writing medical treatises were capable of making, or felt the need to make, that distinction. However, the description of the disease affecting Naples in 1495 *does* seem to be very typical of the progress of syphilis.

No matter how it became established, medical history agrees that the search for a cure for syphilis was a long and confusing one. Because of its exciting relationship to sex, considerable emphasis was put on the morality of syphilis infection. In the 1700s, admission into the French charitable hospitals for the treatment of syphilis was allegedly bookended at entrance and exit by a beating designed to remind the patient of their crimes. It's unclear whether this is completely accurate, but certainly the Hospital of the Saltpeterie in Paris, which catered to indigent and 'fallen' women, was home to a huge number of syphilitics. It's also the site of some of the earliest research into the disease from an epidemiological standpoint; estimates at the time of the French Revolution give the number of those caught destitute or prostitute and interred within its capacious walls at something like 10,000 souls, giving doctors an unprecedented opportunity to observe the effects of the disease. Besides the many physical and neurological symptoms, it led to the identification of the characteristic changes in infant teeth that reveal the presence of congenital syphilis, discussed above. Into this den of iniquity enter one Robert Bunon. Something of a medical adventurer, Bunon went to great lengths to improve his knowledge of dentistry in particular. For reasons that are not entirely given to history, he spent a great deal of time at the Saltpeterie doing experimental things to patients and corpses. His day job was as a celebrity dentist: 'Royal Tooth-Puller to the *Mesdames*' – '*Mesdames*' being the collective noun for the eight daughters of Louis XV of France. However, it is for his considerable work at the Saltpeterie that history remembers him. Alongside influential tracts on oral health, he reports (in rough translation) that the teeth of the infants hide the unfortunate secrets of the mothers and fathers. By this he clearly implies the diagnostic changes in front incisors and first molars present in the teeth of children with congenital syphilis.

While the symptoms may have been clear, the course of the disease was not necessarily so well understood. Lacking insight into the internal workings of syphilis, medicinal texts described the illness as existing in two stages: one with the characteristic sore, and then a subsequent stage where the patient experienced pain in different locations of the body. Many but not all authors seemed to understand that syphilis affected bones in the later stages. Indeed, it would be hard to miss in a very late tertiary-stage patient whose cranial bones have been substantially holed by the disease. The neurological effects of the infection were a little bit more difficult to place, and it certainly did not help that the primary syphilis treatment for hundreds of years was the topical application of mercury.

'A night with Venus, a lifetime with Mercury' – this was the witty euphemistic advice to those who might be thinking about a risky sexual encounter.* Mercury is a heavy metal element, which those of you old enough to remember a pre-digital world may recognise as the silvery liquid that sloshed around in the glass tube of the old-style thermometers. The trouble with mercury is that it's poisonous, and therefore not the best solution to a disease that in all likelihood was going to take its time killing you.† Mercury poisoning can cause neurological symptoms similar to those of syphilis – for example paranoia and an obsession with time-keeping.‡ Toxic levels also cause ulcers, drooling and tooth loss, so it's not surprising that debate raged about its efficaciousness. In 1829 the medical establishment still had come up with no

* And riffs on the origins of the word 'venereal' – yes, it's Venus.

† Albeit horribly.

‡ This last is not supported by clinical study but rather by Disney films, specifically Alice in Wonderland.

better cure than mercury, though there had been warnings about the overuse or overdose of the substance, which had convinced at least some practitioners to give up on it as a treatment entirely. Syphilis was so widespread that literally anything would be done to cure it. According to a survey by historian Kevin Siena, nearly half of the medical ads kept in the British Library dating back to the seventeenth and eighteenth centuries are for syphilis cures. Hospitals in every urban centre were full of the stricken, from Paris's Saltpeterie to London's Royal Free, to the little village of Überlingen and across urban Europe. Syphilis was the product and the opposite of desire; everybody had it, nobody wanted it.

It was the famous anatomist and medical pioneer John Hunter who provided an example of the prosaic, everyday nature of syphilis in the early modern world. In the same way that air travel can be a risk for deep vein thrombosis in some people today, one of the effects of syphilis on the body was a propensity for aneurysm. Coachmen were particularly at risk because they sat on top of jostling coaches, the topside of which would consistently knock against the insides of their knees. With syphilis weakening the circulatory system, a popliteal (that's behind the knee) aneurysm could form with potentially fatal (either the limb went or you did) and certainly painful consequences. Hunter managed to perfect a way to tie off the affected artery, and the troublesome artery as well as a coachman's leg remain on display at the Hunterian Museum at the Royal College of Surgeons in London. Of course, Hunter also thought gonorrhoea caused syphilis, which he worked out by transplanting gonorrhoeic tissue into someone who then developed syphilis, but then he didn't know his experimental subject already had syphilis. No one can be right all the time.

The syphilis epidemic affecting the lustful of the globe by the time of the early modern world of urban cities and nation-states is very well known indeed. What's surprising is that it's very much a mystery how it got to that point, so we must come full circle to consider the question of how we gave ourselves syphilis. The origins of this epidemic have plagued researchers for decades. The most common theories are divided into neat geographic vectors: either it came from the New World, or it was already present in the Old but somehow just took off. A third school holds that syphilis was more or less everywhere. Nearly every historical figure of the European world in the last 500 years has been accused at some point of having syphilis, on more or less reasonable grounds – Mussolini, Nietzsche, Hitler, Lincoln, Beethoven, Schubert, Schopenhauer, everyone mentioned in this chapter … syphilis was nothing if not prolific. In order to evaluate the evidence, we have to look beyond the history books (which, rather prejudicially, are only really interested in white Western people having sex that they may or may not have been supposed to) and turn to a much less prejudicial authority: bioarchaeology. Of course, the bioarchaeology of syphilis has its own problems: as mentioned above, it's not clear if the skeletal lesions alone are enough to distinguish the different subspecies of treponemal infections. Nonetheless, in order to solve the mystery of syphilis's origins, we will have to use what evidence we can.

In 2011 a review of 54 published cases of pre-Columbian syphilis from Europe, Africa and Asia found that none of the cases could be confirmed as definitively syphilitic and definitively pre-dating 1493. Only a scant few of the cases they reviewed were judged to be definitively syphilis and potentially attributable to much older dates, but they dismissed these for lack of clear radiocarbon or other technical dating evidence. If those cases are discounted, this suggests that the New-Worlders are correct in asserting that the epidemic of venereal syphilis arrived in the wake of Columbus's ships. In 2015 congenital syphilis was announced in material from Austria that predated Columbus's return; the notched incisors and mulberry molars were cited as evidence of existing treponematoses. I have, however, taken on something of the attitude of my PhD supervisor, regular Man-About-Teeth Simon Hillson, who finds the consistent diagnoses of specific conditions from the rather undiagnostic evidence of growth disruption in teeth very frustrating. It's difficult to say that any other infection passed to the infant at the stage when the tips of the incisors were forming wouldn't cause exactly the same condition. Happily,* we have come a long way since Bunon, and the ethics boards of modern medical facilities do not allow the sort of experimental research that would directly answer this question.

Of course the problem with archaeological diagnoses is that they are almost always shaded with a degree of doubt until you find the actual arrowhead embedded in the artery.† However, just because there was no conclusive proof doesn't mean that research ceased. Four years later, the results of the mammoth excavation effort at the medieval church of St Mary Spital (not to be confused with the excavations at Spital*fields* – we will get to them in Chapter 12) are finally available, and the crack team of bioarchaeologists associated with the Museum of London's analysis of the material report not just one but seven cases of treponemal disease in decidedly non-tropical London between AD 1200 and 1400. In the part of the cemetery dating to after 1400, they find a big jump in the number of cases, leading them to suggest that while syphilis may have existed in venereal form prior to 1493, it may only have become an epidemic in the post-Columbus world. It's difficult to imagine that the expertise of these bioarchaeologists would lead them to misdiagnose syphilis in a skeleton; I have met or worked with almost all of the authors at some point in my professional life, and the sheer volume of skeletons they have observed in the course of their careers leads me to trust their judgement. And yet, it's very difficult to rest the weight of an entire theory on the diseased bones of just seven individuals and a few statistical models of radiocarbon dates. The authors concede that while yaws was not endemic to cold and rainy England, it may be that these and other cases identified in the same period represent the

* Or not, depending on your level of misanthropy.

† Poor Ötzi. See Chapter 6.

after-effects of contact with the highly connected world of the Eastern Mediterranean – where yaws is endemic –during the Crusades.

The case for treponemal disease on the other side of the Atlantic is much less contested. Characteristic chewing-gum heads and sabre tibias appear across North America and continue all the way through Central America down into Peru and Columbia, dating back at least 4,000 years. While not reviewed as systematically as the evidence for the Old World, it seems that there is reasonable evidence of an antique presence of some sort of treponematosis in the New World. My personal worry with any of the skeletal evidence that does not affect both the skull and the rest of the body is that we ask a lot of our lower legs in palaeopathology. Messed-up tibias are heralded as stigmata of nearly everything: they're either taken as evidence of generic infections, trips and falls associated with farming, crucifixion or potential treponemal infections.* It's interesting to consider early Southeastern US evidence for new reactive bone forming on the shin – which has been previously attributed to the slings and arrows of agriculture – as potentially an indication of treponemal infection, but the diagnosis falls far short of secure. The final difficulty with all of this archaeology is that none of the evidence, New World or Old, has produced much in the way of genetic confirmation of the subspecies responsible.

It doesn't help that *Treponema pallidum* resolutely refuses to be grown in laboratory conditions, making identification of different strands through standard clinical techniques more or less impossible until the advent of genomic sequencing. The unravelling of the treponemal genomes also put paid to one of the three main theories of how Europe came to be invaded by a virulent, sexually transmitted form of a tropical skin disease. Genetics reveal that the disease was not a uniform pathogen, affecting some with pinta and some with syphilis.

---

* They can't all be the Messiah; but neither can they all be naughty boys.

There are appreciable genetic differences between the strains of the disease that affect the body in different ways; syphilis is syphilis because it's the subspecies *Treponema pallidum pallidum*, and not because you ate better as a child or lived somewhere colder.

So, once again, it's pathogen genetics that we turn to in order to extract ourselves from the mess of evidence and theories provided by history and archaeology. In 2008 a study of the treponemal family tree identified a close relationship between the sexually transmitted form of syphilis and the South American strain of yaws. The endemic form of syphilis, the one that isn't transmitted sexually, is indeed a disease of the Old World, it seems; on the family tree it sits further down the branches than the later, more virulent venereal disease. This would indicate that the syphilis epidemic in medieval Europe started as a gift from the indigenous populations of the Americas to the rapacious explorers who began to trickle into contact in the fifteenth century. However, subsequent studies have suggested that this phylogenetic tree has failed to account for the evidence of either the molecular clock of the bacterial genome or the evidence of pathology in the Old World identified far earlier than the fifteenth century.

The genetic split of the venereal form from the rest of the treponemal diseases has been put anywhere between 5,000 and 16,000 years ago; this doesn't fit in well with the narrative of Columbus and his men (and their partners) transforming the disease on its introduction to Europe. Even the most recent genetic studies have a hard time reconciling the dates and places of genetic splits between the various subspecies of *Treponema pallidum*. Perhaps, really, everyone is wrong on the causes of the virulent outbreak of syphilis in medieval Europe. An interesting suggestion comes from the evidence that the treponemal disease that affects baboons in Africa is closely related to the subspecies that causes yaws. Perhaps Columbus was a red herring, while two disparate, long-lost treponemal cousins performed a coordinated pincer movement to infect medieval Europe, as the same nations that sent Columbus west started pushing further and further south into Africa in search of the exploitable, valuable and exotic. Whether the true genetic origins of the syphilis epidemic lie in the New World or in Africa, where the European crisis in bullion led to a determined push for exploration in the decades before Columbus's voyage, it seems clear that the rapacious economies of late medieval European cities, with their mobile armies and clever ships, brought home more than they bargained for.

Why do we care about syphilis? Obviously, there is fun to be had with euphemism and *double entendres*, but that's not quite enough justification for this many thousands of words.* Syphilis is interesting because it's a truly horrific disease, morally charged, and we still don't understand the circumstances in which it arose. The modern world still experiences plagues, and the ravages of HIV/AIDS, while not identical to the syphilis epidemic, offer sobering parallels. Understanding the circumstances in which this disease, of obvious antiquity but in uncertain form, prospered as the primary preoccupation of priapics† is critical to understanding the processes that are still at work in our society.

One of the first attempts to reconcile the theoretical origins of syphilis and evolutionary theory comes from E. H. Hudson, who saw the treponemal diseases as a progression of the same disease adapting to new circumstances, and this is where cities come back into the picture. In his telling, yaws the hot, wet treponematosis of Africa, migrated to the colder, drier Near East and became endemic syphilis. The disease made its next big leap to venereal transmission with the invention of cities. Hudson cites clothes as one potential contributing factor. It's true that the cities where syphilis first made its mark are full of people wearing clothes, and Hudson argues that clothes, particularly on children, limit the skin-to-skin contact that would normally spread treponemal infection.

---

* Even for me.

† See? Awful.

Those encountering it would either gain some immunity and never develop the full-blown condition, or would develop whichever non-venereal form of treponemal disease they managed to acquire. Hudson argues that this might be impetus enough to coax an endemic skin infection to finding its way back into the body by hiding out *under* the clothes – and making the jump to venereal transmission. Judging by the murals on the walls of Maya cities,* however, partial nakedness is no barrier to urban life. Hudson's theory is only viable in a population where there was no stand-by immunity from childhood messing-about and no potential for skin-to-skin transmission in adulthood. The argument that naked Neolithic children provided a blanket of childhood exposure to treponemal disease that conferred immunity from the venereal infection is hard to square with the reality of climate in the Near East, which, despite what you might imagine, gets very, very cold. Clothing is not entirely optional through some of the seasons, even in the Neolithic, despite what so-called 'educational' murals full of topless, pensive, fire-tending women would suggest.

In a further characterisation of the syphilis-friendly urban environment, Hudson accuses cities of simultaneously narrowing women's contributions to the world and inventing prostitution, and therefore giving syphilis a venereal haven. But of all his arguments, the most unbelievable is the one that declares cities to be so hygienic in comparison to villages and hunter-gatherer life that endemic syphilis dies out. He argues that it was cities that caused syphilis – not because they were dirty, but surprisingly, because they were *clean*. This is an extraordinary accusation to make of pre-modern (and many modern) cities, even in 1965, and one I think Chapter 13 will put paid to. The argument that cities provided options for soap and separate beds seems to fly in the face of what we know about disease spreading in cities and urban architecture

* And even European capitals in summertime.

and among the disenfranchised that urban social hierarchies relegate to slum living and health inequalities. It also skips rather quickly over the cases of treponemal infections observed in Maya contexts; they had cities too, albeit often more spread-out ones.

Of course, Hudson is not the only one to point to the rise of urban life as a factor in the spread of disease. Other researchers have also implicated cities by suggesting that the earliest settlements may have offered greater opportunities for sexual relationships due to the increased number of potential partners in a particular geographic radius, but to my mind this is to rather miss the considerable evidence for similarly elaborate sexual mores among various non-settled peoples.*

I personally rather like eminent archaeologist Shelly Saunders' comment on the task of understanding the complex picture of treponemal diseases: 'The novice syphilologist, sifting through a voluminous literature on the origin of the disease, must feel like the wind-blown sapling bending to each successive argument.' However tumultuous the process, there are nonetheless a few clear points to cling to. Treponemal disease has a long history with our species, with genetic and phylogenetic evidence suggesting that the separate subspecies of the bacteria are very much different. How they relate may eventually be answered either by future successes in retrieving archaeological pathogen DNA or by concentrated work collecting further evidence, particularly from Africa. There is some rationale for considering an African source of treponemal disease, which then makes an entry into Europe around the fifteenth century. In the 1400s European exploration expanded into contact with sub-Saharan Africa through sea voyages and even overland routes. The impetus

* Wife-swapping, polygymy, polyandry, temporary marriages … ethnography is eye-opening.

for this may have been the continent's first real capital crisis, the 'Great Bullion Famine' that saw raw metal production plummet, taking the nascent capital economies of southern Europe with it. European exploration into Africa was designed to subvert Arab control of the gold market and the rest of the luxury trade from the East; contact with these new disease environments has been very poorly investigated archaeologically. It would be very interesting to see what bones can tell us about what fifteenth-century African trade – in raw materials and in diseases – looks like from an African perspective.

Whether syphilis arrived in Europe with Columbus or Crusades, or was simply always there, it's clear that there was an epidemic of fatal venereal disease in the fifteenth century that seems to have lasted until the development of penicillin some 400 years later. The nature of syphilis is relentlessly urban, whether its root origin was in urban living or not. It spread between the connected cities of the newly global world, reaching China and Japan only a few decades after making its debut in Spain. It is the key components of urban connection that allow epidemics to flourish in the way that syphilis did: population density and the presence of social hierarchies that leave the elites at the top no option but to churn desperately through luxury items such as exotic trade goods in order to prove their status; those luxury goods and new diseases travel along the same roads. Perhaps the Black Death in fact laid the carpet out for its successor by prompting the social and economic upheavals that freed European populations from being locked to their land under feudalism. It's easy to see how the increased professionalisation and wage economies of the later medieval world might have increased the opportunities for both mobility and transmission of sexual disease. That same wage labour is also potentially a factor in the changing nature of European economies, ending up with everyone running out of exchangeable currency at exactly the wrong moment and needing to set off for Africa and the New World to find more gold. I would never go so far, however, as to accuse the cities of this new, glorious, globally connected world of being *healthy* or *clean* – as the very last example of the horrors visited upon our poor mortal forms by urban living described in the final two chapters will demonstrate.

CHAPTER TWELVE

Take This Job and Shove It

Having reached a point in this book where we have arrived at a thoroughly modern age (well, the start of one anyways), it is incumbent upon me to take us right back to the beginning. Like the worst sort of history teacher, handing out a pop quiz right before summer break, I'm going to ask you to remember what we talked about in Chapter 2. No, not the Bedouin kids laughing as they fixed the British School Toyota while the great minds of Neolithic archaeology stood idly by – actually, wait, kind of, yes. Because this chapter is about work. It's about moving from the eloquent bones of Abu Hureyra to the vicious scars of factory life. Work is what we do; it is determined by the world we live in, and it leaves its mark on our bones and teeth. What we have to figure out, really, is how urban life has changed work for our species,* and what our bones say about those changes, for good or ill.

The modern reader may well have personal knowledge of the crushing grind of working life in the city.† Until we achieve the 'Fully Automated Luxury Communism' envisioned by the Scottish school of science fiction led by the late and much-lamented Iain Banks, work continues to occupy the vast majority of our otherwise enjoyable‡ lives.

---

* And, more specifically, how I ended up locked for years in a medium-sized, asbestos-lined box full of things that go beep and a microscope that goes 'sneesneesneesneesnick'.

† The modern reader may equally fervently wish for the opportunity to know the crushing grind of employment, particularly if they have an archaeology degree.

‡ Unless, of course, you work in archaeology, which is terrifically interesting but rarely clears the payment-for-labour definition of 'work'. You can't really win this game.

While labour is not the exclusive preserve of urban life (recall the Hadza hunters shimmying up trees to brave angry bees in order to get a bit of honey), there are certain types of labour that exist in the urban world and certain patterns of work that leave their mark on our bodies. For the moment we will leave aside the very modern phenomenon associated with the digital revolution, despite the extreme relevance of repetitive strain injuries to the completion of this volume, and we will consider the idea of what work has really meant as we have made the slow transition to an urban world.

Farming, at first glance, seems like an easy job. The whole thing revolves around plants that, once planted, grow whether you're watching them or not. Of course they don't, actually: birds eat the grain, more terrestrial animals nibble the stalks, there's never enough water, or perhaps there's a late frost and you've got to give up and do it all over again. Agriculture in its earliest form was still work, for the vast majority of people. Intentional cultivation means planting, which also means having soil in which to plant – the soil can't be too barren, likely to run down the side of a hill, or under a metric tonne of jungle. Traditional methods of clearing fields include cutting or burning down overgrowth, or you could go the opposite direction and build up terraces on steep slopes that would catch the soil before it goes tumbling downhill. It's one thing to contemplate fields of wheat from a car window, and another thing entirely to get grains from seed to Seder at the dawn of agriculture.

In the summer of 2012 I set off to the site of Aşıklı Höyük. It's set on the very edges of Cappadocia, the famously beautiful landscape of hills, caves and chimney houses that attracts millions of tourists. Traditionally, archaeology requires that you suffer in the doing, but since I was on a flying visit, I'd allowed myself the luxury of a rental car. Cruising down the interminably under-expansion highway from Ankara to Aksaray, I passed the Great Salt Lake and of course endless fields of wheat. Anatolia could give Kansas a run for its money on a good day for rolling fields of grain; Turkey exports some

4 million tonnes of the stuff every year.* Turning off into the tiny village of Kızılkaya, I was about to come face to face with the origins of that enormous enterprise. The Aşıklı Research Project has long been a proponent of experimental archaeology, building an entire Neolithic neighbourhood of varying types of mud brick using techniques borrowed from the vernacular architecture of the surrounding village. They have even let one of these structures rot peacefully in the hopes that it will reveal key insights into the decay process of the 10,000-year-old mud-brick houses of the archaeological site. Just off to the side of this reconstructed village is a small patch of ground, about 2 by 2 metres (7 by 7 feet), which when I visited in late summer was a vibrant sea of green, bar the solitary figure who stood in the middle, master of a multi-year experiment to see how the green things grow.

This was Müge Ergun, an archaeobotany student and archaeologist. She'd decided to see what farming at Aşıklı might *actually* be like by recreating a semi-managed field, akin to what the very first farmers might have done, using the wild strain of einkorn wheat. She had also created the tools she would need to harvest her crop based on examples found on site: razor-sharp scythes made of shining obsidian blades hafted onto a crescent of wood with the aid of sticky bitumen. The obsidian at Aşıklı Höyük comes from the mountain of Hasan Dağ, one of the key sources for tools with edges sharp enough to hunt … and to farm. Müge was kind enough to let me into her patch, and even let me examine her Neolithic scythe. It had wickedly sharp edges, but was annoyingly short-handled: appropriate probably for the small-scale harvesting Müge was planning on doing, but not quite Grim Reaper territory. She planned to tackle a bit of her patch that day, and I fully intended to stick around to watch. After the first few passes, however, I realised two fundamental problems with farming: first, it's actually quite tedious to watch, let alone

* Okay, Kansas exports around 10 million. But that's just out of Kansas.

do. And second, it can*not* be good for your back. I quietly went back to the main house to bother someone else.

Farming lifestyles, it seems, are not particularly kind to our bodies – or our backs. Looking at the wicked edge on that scythe, it suddenly even seemed mildly dangerous. The danger was not all from sickle-based accidents in the Neolithic, of course; work itself left its marks on our bones. We can catch glimpses of what this work would have entailed through studies of robusticity and muscle activity, two very different but complimentary bioarchaeological techniques that tell the story of our bodies' habitual movements introduced in Chapter 2. We can also see the wear, tear and accidental damage that transitions into agricultural and eventually urban life have caused. Together, these make up the bioarchaeology of human activity, which is a surprisingly contentious field given that we know people must have been doing *something* to leave all that archaeology behind.

We discussed the case of the build-up of the muscle activity markers way back in Chapter 2: this is where the bits of bone that hold onto muscle essentially get a lot more rugged with greater activity, building up a greater surface to hold bigger muscles. Generally, hunter-gatherers are seen to be more muscle-y than agriculturalists, with more evidence of muscle attachments across the body. One of the most comprehensive studies of changes in human activity patterns is presented by Jane Peterson, who looked at the patterns of musculoskeletal stress markers (MSMs) across a wide range of sites in the Southern Levant. MSMs are the kinds of tracks left on bone by habitual activity and by comparing evidence of muscle use across societies more or less firmly embedded in the practice of agriculture and animal management, Peterson came to the conclusion that in the Neolithic male activity markers seemed to switch from unilateral to bilateral, while for females bilateral markers remained the norm. This, as Peterson points out, is very difficult to explain in terms of specific activities, but it seems that men's activities may have changed from more one-sided efforts to those involving symmetrical strain; we might think about the effort that goes into using certain types of hunting weapons versus two-handed grain grinding as posited by Theya Molleson for the women of Abu Hureyra.

Transitions across time are difficult to capture, but a few studies have tried to demonstrate the changing nature of work in periods of transition. Peterson's study of the Levant, for instance, included the transition to the Bronze Age, a point in time where, if there were not always cities *per se*, there were certainly more densely settled populations. At this point she sees another transition in activity markers, with females demonstrating a higher prevalence of one-sided activity and more upper-body involvement than males of the same period. Activity markers in Italy have also been studied in both the Neolithic-to-Bronze Age transition and across the later metal ages. The examination of signs of muscle attachment from a necropolis dating to 10 centuries either side of AD 0 suggested to a group of researchers that men manfully rode on horseback and used weapons,* while women did women things. You can begin to see the problem here. Muscle use in the past can be attributed to anything you want it to. If you expect manly men doing manly things, and your personal cultural experiences have already decided what those things are, then it's hardly a surprise to find your biases confirmed in the skeletal record. Determining activity from markers of activity is not quite as bad as reading tea leaves,† but surely at some point one must question interpretations such as: 'The females may have been submitted to greater mechanical stress especially while carrying a baby in the left arm and simultaneously carrying on some sort of domestic labour with the right arm.' Do all females have babies of carrying weights at all times? What about the left-handers? How long have you ever tried holding a baby and doing anything more complicated than

---

* This one I will allow is possible; they found the horse, the armour and the weapons. However, until it's confirmed to be a consistent finding for all men buried with their medieval fighting kit, the jury remains firmly out. See the discussion in Chapter 8.

† You are spared the effort of making tea, for one.

checking Twitter? This is the worst sort of just-so story, a preconceived narrative with the social norms of the modern world projected back on evidence that can't quite be taken at face value.

There is more than one way of making a skeleton talk, however. The second technique for capturing patterns of activity from the human skeleton is something that has really only come into its own with the advent of, of all things, big-money professional sports. In conjunction with advances in medical imaging, science has begun to look at the shape of bones, particularly the limbs, as an overall indication of biomechanical strain in life. Bone, as I've repeatedly pointed out, remodels. It remodels when faced with infection, when broken, and just in the normal process of growth. Where there is mechanical pressure, bones tend to add more bulk through a feedback loop of biomechanical signals; this is why unused limbs atrophy. Athletes, on the other hand, build up the bone they need; this is what several pioneering studies of the skeletons of still-living sportspeople revealed in the magical eye of the CT scan. The overall robusticity of bone can be judged in a few different ways, the choice of which has generally followed the availability of new technology. You could run a measuring tape around a bone; you could look at how bulky the bone is in general by measuring the hard outside area alongside whatever space is left inside of a cross section of bone using X-rays or CT scans; the outer surface of bone could be 3D-scanned to make a shape profile at key points for comparisons; you could even go microscopic and adopt the sophisticated internal bone structure 3D-modelling techniques pioneered for the study of (modern) osteoporosis.

As you might imagine, the earliest investigations of bone mechanics were fairly limited by the difficulty of assessing how much bone was bone. It only took seven years after the invention of the X-ray for anthropologists to realise the advantage of seeing inside the bone, and the discovery of the Krapina Neanderthal remains provided pioneering Croatian anthropologists with an excuse to irradiate things in the name of science in 1902. It was considerably later that the science of human skeletons got around to identifying the mechanical properties of bones as problems of geometry, but cross-sectional geometry of long bones has been a standard measure of bone robusticity since. Differences in the use of the limbs do seem to link to differences in the shape of a cross section cut out of one of the major bones, either through X-ray or with an actual saw. Sports with a high mobility component, *e.g.* lots of running around, seem to increase the triangular nature of the shin, whereas less active lower legs are more circular in cross section. This makes sense from a mechanical perspective: the strain on the bone is being taken up by the chunky forward projection of the tibia.

Archaeological applications of this theory have been used, just like muscle markings, to recreate patterns of activity in the past, and a very similar pattern emerges. In the Neolithic, bone loading seems to decrease, something associated with the lower mobility of farmers. Anthropologist Jay Stock of the University of Cambridge has comprehensively studied the relative robusticity of bones across the long stretch of human history beginning with the Epipalaeolithic of the Levant (before our Natufians) and through a very long series of early agriculturalist remains in Central Europe. His evidence shows that a reduction in bone loading – and therefore a reduction in activity – appears with the transition to agriculture. This is why, in Chapter 1, I accused farmers of having skinny legs.

Not all farmers experienced the same transition, however; the maize agriculturalists of the Mississippian culture seem to have increased the strength of their bones in the Neolithic, particularly the women. Other notable exceptions are early farmers from Italy who seem to have foolishly chosen to live in mountainous regions and therefore kept up the pressure on their legs. However, by the Late Bronze Age, Central European males had lost much of their lower-limb strength, suggesting a further reduction in mobility. Even our Italians of the Middle Ages show sexual dimorphic labour patterns that we might normally expect in a more agricultural setting, where women's and men's work is supposed to diverge. In a study of two more or less pastoral or settled groups,[*] which while over-interpreted in the case of the muscle markings, may link to real lived experiences, men showed more one-sided bone robusticity in the upper body, while women developed both sides, and in the slightly more settled group lower-limb robusticity was reduced.

On the other side of the Atlantic, the bodies of the Classic Maya at their most urban, around AD 200 to 900, show characteristic changes in long bone loading that signal changes in the social order, not just in subsistence; remains from Xcambo show that becoming an administrative capital generally decreased the workload on long bones. The process of becoming a city, however, may have required more effort: a recent study at Tell Brak in modern-day Syria has suggested that lower-limb robusticity between the period of becoming a city (the late Chalcolithic, some 6,000 years ago) and of existing as a fully-fledged city (the Early Bronze Age, about 5,000 years ago and contemporary with my work at the site of Başur Höyük) also decreased. The authors suggest that the decrease, seen really only in potentially male skeletons, might indicate that the urbanising phase of the city still required quite a lot of mobility.

Of course, the work of life does not just build up bone. It also takes it away, as anyone over 35 who has ever tried to rise too quickly from a long-held recumbent position will be well aware. Bone does not exist in isolation in the skeleton; it is supported and held together by connective tissue, and the moving ends between bones are encapsulated in joints. Joints have the misfortune of being absolutely critical to the use of the skeleton, while simultaneously being built of much less stern stuff. Habitual wear and tear[†] causes changes in the joint capsule that can eventually lead to a collapse of normal

---

[*] Contributing to this study was eminent Italian archaeologist Alfredo Coppa, an exceptional host as well as archaeologist. I'm not going to forget whizzing around Rome (the Coliseum! Gianicolo!) on his scooter anytime soon.

[†] 'Living'.

function; this is degenerative joint disease. Degenerative joint disease is the name given for a whole suite of changes affecting the joint, from compression onwards, and most people will be more familiar with the changes that occur once the problem gets into the bone: osteoarthritis. Looking for activity-related changes by surveying a skeleton for arthritis seems intuitive, but it turns out to be remarkably complicated. There are several confounding factors, primary of which is age. Age bequeaths wisdom and eventually subsidised bus passes, but some time before that your body begins to respond to the pressures applied by movement, lifestyle and even gravity.* Added to the effects of sex, population, metabolism and factors we have not quite got a handle on, this makes the study of degenerative joint disease in archaeology equally promising and frustrating.

Osteoarthritis sounds like an infection† but is actually not, strictly speaking, an inflammatory response; the bone is not infected and does not react in the same way it does in infectious or even metabolic diseases. Stress put on a joint affects the articulating ends of the bones that meet within it. The bone can respond by forming osteophytes, which are little bony spicules that form a lip around the edge of the articulation. If the joint space and tissue totally collapse between the two bones involved, they can rub directly on each other. This creates a particular effect called 'eburnation'; essentially, the bone ends polish one another. Bioarchaeologists have long been interested in using the map of degeneration provided by skeletal evidence of osteoarthritis to reconstruct the world of activity in the past, but the relationship between osteoarthritis and activity is in no way perfect. A variety of factors including age, hormones, body weight and even population variation go into creating osteoarthritis. Additionally, much of what we

---

* NASA informs us that astronauts can gain up to 5 centimetres (2 inches) in height in space. While in day-to-day life avoiding gravity may not make up for the concomitant loss of bone strength from reduced loading, the view is probably worth it.

† '-itis' is the common suffix for infections.

know about joint stress and arthritic changes in living people, who can[*] report their activities accurately, comes to us from radiographs and CT scans that measure the space in between two bone ends and therefore identify changes far before they start to affect the skeleton. So really, while looking at osteoarthritis to reconstruct activity is an 'alluring prospect',[†] we must also take any interpretations with a degree of caution.

Patterns of osteoarthritis in hunter-gatherers largely suggest a heavy and diverse workload; the evidence of degeneration is mostly in line with the evidence of new bone formation for carrying muscles, discussed above. A somewhat strange study done in the mid-1940s purported to show that the Sadlermiut Inuit people of Hudson Bay, Canada had wear and tear in their skeletons resulting from very specific activities. Men's upper arms were bilaterally built up due to kayak paddling, and compressive fractures to arthritic vertebrae were 'snowmobiler's back', due to sledding; women's jaws were arthritic because of the cultural expectations of preparing skins with their teeth.[‡] The differences between the arthritic changes in the hunter-gatherer and in the farmer do not seem to be very pronounced across a wide spectrum of sites; tabulating cases of arthritic knees, elbows, wrists, *etc.* in farming versus foraging groups turns up very little statistical difference. Bioarchaeologist Robert Jurmain has studied arthritic changes related to activity for decades, and has compiled several such examples. In some groups of hunter-gatherers, there is evidence of arthritis in the wrists and hands that is not present in most agricultural groups, but beyond

[*] But do not always.

[†] In the words of Tony Waldron, who has a proud history of scepticism of too-neat bioarchaeological stories.

[‡] According to several accounts, Inuit wives were in the past expected to chew their husband's boots to appropriate comfort levels when called upon. What is remarkable is that this is possibly the least complicated aspect of Inuit cultural rules on marriage; if you ever want to be really perplexed, try understanding the system of wife exchange.

this occasional finding, it's difficult to pin down changes in activity patterns with any certainty.

The problem with translating wear and tear into patterns of activity in the past is that the relationship between using a joint and wearing down a joint to the point where we can see changes on the skeleton is not perfect. Sheep have been walked on cement and rabbits on treadmills in efforts to reproduce in a laboratory the arthritic changes we see in skeletons; treadmills are apparently more forgiving than sidewalks.* There are some positive correlations identified in modern medicine, however, and these tend to be from either injury-related changes to the joint or very specific types of activity.

We can take, for example, the very famous case of the excavation of the early modern period Huguenot cemetery at Spitalfields, London. What seems to have been every archaeologist able to hold a trowel in London in the mid-1980s worked on the enormous Spitalfields Project, excavating Christ Church Spitalfields in East London. The dig, and subsequent work in labs and archives, have provided an extraordinary treasure trove of information. The people who had come to be buried in the eighteenth-century crypt of Christ Church Spitalfields had the excellent foresight to be relatively middle class and literate, so that some of the doings of their lives were deemed worthy of report in the burgeoning printed news trade, in legal documents surrounding property and estate issues, in records of occupational guild memberships and, best of all, in elaborate curling script in metal plates attached to their impermeable lead coffins.†

* Something you could have asked any runner.

† Lead coffins are a blessing and a curse; everyone is always terrified of opening them in case bits of anthrax or smallpox leap out (whether this is an actual scientific possibility or not). They can also cause staining in the bones themselves because of the fluid build-up inside; many of the bones from the Spitalfields skeletal collection are a startling shade of bright pink.

For many of the interments whose names and birthdays were carefully recorded on their coffins, full biographies can be cobbled together from parish birth record to last will and testament. This allowed bioarchaeologists to make comparisons between the signs of activity on actual skeletons and the reported or predicted life experiences from known patterns of occupation for this group of relatively well-off men and women, many of whom were established tradespeople with the insurance documents to prove it. The silk industry accounted for the major part of employment, a trade carried over by the Huguenots in their escape from France, and which made the fortunes of several individuals. Of the 29 identified silk weavers, who might have spent their entire working careers making exactly the kind of repetitive motions people have theorised are associated with the formation of bony arthritic changes, only three had hands showing the characteristics of osteoarthritis. This is rather a blow to the idea that we can deduce activity or even occupation from specific patterns of osteoarthritis.

It's very difficult to reconcile this rather total lack of correlation with the work of early pioneers in the field, who did find the arthritic changes they were looking for when applied to pre-historic populations for which behaviour could only ever be inferred, and never known. There are a few suggestions as to why patterns of osteoarthritis do not reflect what we know to be true: humans in the past did different activities, habitually. Bony arthritis seems less common in the warmer parts of the world, but it appears more often in the obese and the elderly: a pattern that may reflect a complicated interplay of both activity and population-level genetic variation in susceptibility to the factors that influence the changes in joints and bone. Of course it's these factors, which bioarchaeologists would like to be habitual activity, that are in question. One answer, put forward in clinical research, is that the not-really-a-vitamin vitamin D might have a role to play. Science is slowly considering evidence that vitamin D has its proto-hormonal fingers in quite a lot of our structural selves, and that the effects it has on bone growth may extend to the bony spicules of encroaching osteoarthritis. It was work on the Inuit groups of North America that first spurred interest in osteoarthritis as a palaeopathological marker of activity; wouldn't it be nice to find out that it's in the frequently vitamin D-deficient groups from those high northern latitudes that the technique is actually applicable?

One of the earliest examples of the danger of work comes from our kissing cousins, the Neanderthals. The famous Shanidar Caves in northwestern Iraq hid the remains of at least 10 Neanderthals for tens of thousands of years, many of them with fairly gnarly skeletal injuries. There are poorly healed fractures that would have led to wasted, useless arms; dings and dents to the skulls; and cracked ribs aplenty. The bodies of the late Stone Age are liberally dressed with traumatic scars, particularly to the head and neck. Lee Berger, famed discoverer of the new species of hominid *Homo naledi* in the Rising Star cave system of South Africa,* and the esteemed palaeoanthropologist Erik Trinkaus co-wrote a paper in the mid-1990s searching for comparable pathologies and came up with a rather unexpected modern correlate for the injuries they saw: rodeo riders. They theorised that the Neanderthal 'job' of hunting down enormous, pissed-off animals with horns and a serious weight advantage using only close-range weaponry like spears was likely to end up in someone (or maybe everyone) getting hurt. As a story, it was excellent – I certainly passed that question on my undergraduate exam with flying colours. But was work before civilisation really that dangerous?

Trinkaus has revisited this issue in the intervening years and come to a different conclusion. He argues that the rodeo-riding injuries Neanderthals suffered are pretty much the same ones that our *sapiens* ancestors living at the same time picked up. Also, there's a bit more evidence now that

* And boy, if you ever want your stomach to churn at the thought of work, watch the video of the cavers going into the *Homo naledi* cavern … through a 28-centimetre (11-inch) gap. By dislocating their shoulders. On purpose. Science can be intense.

Neanderthals were less inclined to stand around poking enraged animals with sticks and a bit more on the distance-throwing end of technology, so close-quarter hunting can't have been the only reason for that pattern. It may actually be the case that life in general was full of bumps, falls, kicks and walking-into-things, and what we see written into the bones of the Shanidar Neanderthals is the legacy of living in a world without proper street lighting or road safety cones.

The pattern of accidents that befall our species does seem to shift with the Neolithic transition. Much of the trauma we talked about in Chapter 6 that occurs in the periods before sedentism might be attributable to accidents – not necessarily *intentional* violence,* but to the risks of an active, food-pursuing lifestyle. In the Neolithic, new ways of living occasioned new risks. One of the unexpected findings is that houses themselves seem to cause a great deal of damage. We saw in Chapter 7 that a bioarchaeological survey of the injury patterns of the Ancestral Puebloan culture showed a large number of both cranial fractures (which may or may not be related to levels of violence exacerbated by pressures on community survival) and, intriguingly, compressed vertebral fractures. The location of a compressed fracture in the spine gives a clue as to cause; we've talked about the pathological conditions that can collapse spines in Chapter 9, but in the absence of evidence of infection, what are we to make of such injuries? While the most common cause of compression fractures of the vertebrae today is the thoroughly modern road accident, followed rather closely by slips and falls, other trauma can also squish your spine. Falling, it seems, is liable to crunch your vertebrae, through force being transmitted directly down through the column and, if strong enough, collapsing the cushions between the vertebrae or even the bones themselves. One thing we know for certain because of the excellent preservation of the arid domain of the Ancestral Puebloan groups is that their impressive cliff-face dwellings,

* Though anyone who has walked through a dimly lit living room will know that coffee tables are inherently malicious.

honeycombed into rock undercrofts in multi-tiered mud brick, were accessible by ladder. As any child who has ever had the top bunk knows, ladders are an invitation to fall from a height. You don't necessarily land on your behind, either; blows to the head and typical forearm fractures like the Colles' fracture might also be the result of falls. It may be that even the type of housing we constructed to help us live year-round in the same place managed to break our bones.

Farming has been held to result in higher rates of damage to the lower leg, evident to bioarchaeologists through a combination of fractures and general inflammatory response; in a group of First Nations Iroquois from Ontario, these numbers were elevated with the transition to farming. While I'm not entirely sure of the mechanism suggested for these two types of accidental injuries* (Theya Molleson has suggested that they might relate to the potential of traditional plough furrows to trap lower limbs during falls), they do seem to appear more commonly in agricultural contexts than among hunter-gatherers. There seems to be a pattern of wear and tear related to the flexion of the spine used in manual harvesting methods that accompanies agriculture as well as falling off the face of your vertical cliff village; vertebral fractures in the mid-spine of two women from Neolithic Jiangzhai in central-eastern China have also been suggested as signs of a life of agricultural labour. There is a longstanding joke in bioarchaeology, particularly among those active in fieldwork in boggy, terrible soils like London clay,† that the only modern profession to equal farming for the strain and torsion of hips and spine is … archaeology. So far, no

* Though I can imagine a situation with an over-exuberant sweep of a long-handled scythe a bit too close to the body that ends in tears.

† Archaeologists who have not worked in the British commercial sector may be unfamiliar with the UK habit of moving literally tonnes of heavy, wet, boot-sucking clayey soils with a short-handled shovel. It's murder on your back if you do it wrong. Or, even if you do it right, if you do it long enough.

archaeologists have volunteered themselves for a scientific test of this suggestion, but I'm sure it's only a matter of time.

Accidents linked specifically to occupations are very difficult to identify in the past, a time before occupations were the specialities they are in the urban world. While still few and far between, we have some examples of what sort of accidents urban working precipitated, but even more interestingly, we have the suggestion of the accidents they did *not*. Going back to the Huguenots of Spitalfields, Tony Waldron identified a much lower overall rate of broken bones in the urbane world of the eighteenth-century silk trade than in the towns of Roman Britain. Perhaps life wrapped up in silk was just safer, but there seems to be some suggestion of decreasing rates of fracture from the Roman period, so increasing urbanism (if you can call it that, given that the Anglo-Saxons were really very villagey) in England may have actually decreased the sorts of accidents that break bones. Charlotte Roberts and Margaret Cox have carried out a comprehensive survey of bioarchaeological evidence of every kind in England, from prehistory to the present day, and they attribute many of the fractures unrelated to violence as potentially the result of agricultural work – the aforementioned furrow-falling, or, as they suggest, encounters with carts and the things that pull them.

It's this last danger that is most familiar to modern lives; road accidents then as now were a real source of danger. Urban living, with its crowded, narrow streets, resulted in traffic conditions we might expect today; like the three-hour traffic jam that occurred on London Bridge on 21 April 1749, set off by a bit of Handel and some slightly out-of-control fireworks. They also resulted in a horrific number of traffic accidents. Two of our Spitalfields folk had fractures to the hip that in modern forensic practice would be most likely to result from a car crash; the bioarchaeologists who identified them suspect a similar cause was at work in the eighteenth century. Urban living, it seems, freed us from some sorts of accidents, but left plenty of room for us to invent new ones.

Having started with the Marxist revolutions of Childe, we come to a revolution universally agreed upon: the Industrial Revolution, which allowed the making of things in quantities previously unimagined for consumers previously unlooked for. Unlike the fad industries of the pre-industrial world,* industrialisation was a process applied to many different realms of working life, and therefore affected a concomitantly large number of workers. The Industrial Revolution was full of jobs, but many of those jobs were designed with the productivity of industry – and not the happiness of the worker – in mind. Constrained repetitive tasks in 'assembly line'-type factories might lead to identifiable patterns of muscle attachment, bone robusticity and degenerative joint disease that are far more specific, and therefore far more traceable, than those carried out in the comparatively varied agricultural workday. However, we have seen how well those studies have turned out: identifying either very vague patterns of muscle build-up or loss or the untrustworthy evidence of osteoarthritis.

An exception to this frustrating pattern, however, can be made with the invention of particularly unhealthy labour. Certain highly specific occupations, defined and delineated by the capitalist economies of the Industrial Revolution, led to exposure to risks that, eventually, would spark the 'health and safety' culture of the modern workplace that tabloid newspapers so love to deride. The sneering attitude towards workplace protection rather misses the fact that, in the past, work could kill you. Jobs like mining, sponge diving and lion taming are all recognised, now as then, as risky propositions. But what if you didn't know your job was killing you? The character of the Mad Hatter in Lewis Carroll's *Alice in Wonderland* causes mirth and some bemusement today. It seems an odd choice, making a scatterbrained character of uncertain mental facility be a

* Dutch tulip, anyone? Or perhaps some Roman fish sauce worth its weight in gold? Maybe you'd prefer a nice obsidian core from the Hasan Dağ mountain in Turkey?

hatter. But step back a hundred years and the connections in people's minds would be immediate: hats were quite commonly made of felt, and felt was best made by mercury. Mercury is not a thing you should spend a lot of time handling, breathing or taking as medicine (as discussed in the previous chapter), but it does do wonders for separating out animal fur, something that only seems to have been providentially discovered by replacing the traditionally used camel urine with human urine, some of which was tainted with mercury due to the popularity of mercury cures for syphilis at the time. Either way, workers who came into close contact with it frequently suffered the symptoms of mercury poisoning, and ended up 'mad as a hatter'.

Many readers will be aware of the story of the little match girl. A slightly Danish version of a morality tale written by Hans Christian Andersen, the little match girl is an impoverished waif reduced to selling matches on the street.* For reasons that are suitably tragic, she dies in a snowbank on some dirty city street.† Of course, while the world wept at her plight, very few were aware of the fate of the men who made the matches themselves. Matches were once made of phosphorus, which burns brightly if somewhat irregularly. They were made in factories with huge vats of the stuff, and the little stick parts of the matches would have to be guided into the phosphorus by a worker. These workers would spend hours, even years, inhaling phosphorus. First it would creep into the mouth and the soft susceptible tissues of the gums, where gangrene would set in. From there, the condition would progress by blackening and killing all the tissue in its path – necrotising the bone so that it rotted away

---

* There are uncomfortable parallels with the children in many countries who sell bottles of water, sticks of gum or packs of tissues to cars parked at stoplights or along train lines. The world has not moved on so very far.

† This depends on the version. My particular favourite being from 1986, featuring Twiggy, Roger Daltrey and a friend who cringes every time I bring it up.

inside the matchmaker's mouths. The dead bone could no longer hold the teeth or even itself together. There are several examples of 'phossy jaw' in medical museums; legend holds that they still glow at night when the lights are off.

At some point we also have to consider the power of cities to draw in labour, and to exploit it. This can be an unconsidered consequence of the availability or promise of employment, or it can be an institutionalised programme of forced labour. Quite a lot of it happens outside cities, but it takes an organised power structure to repurpose human labour from the kind that serves the personal and communal to the kind that builds monolithic structures for us to take selfies in front of. This is actually an interesting, and recently contested, point – until fairly recently, pretty much any Big Thing built by human hands has been thought to be spurred on by hierarchical, complex societies: think lofty pharaohs and giant pyramids. To get to hierarchy and complexity, the story went, you had to sit down and accumulate surpluses, something only the major urban polities of the world were capable of. This rather dismisses the potential for collective building efforts, and flies right in the face of the giant symbolic pillars of places like Göbekli Tepe, which predate any sort of sedentary society in the region.*

So, despite it being *possible* to build big things with voluntary, collective labour, we will consider here the more common reality: it usually takes a densely settled state to pull in the kind of labour force you need for really big projects. Sometimes, it's for projects perceived critical to the survival or success of the polity; sometimes it's just to make something big. The strange animal wars of the Cultural Revolution period in China are an example of a state co-opting every member of society to a particular labour; in the famine years of the early 1950s citizens were commanded to go out into the fields and kill sparrows, which preyed on the increasingly precious grain. This resulted in a lopsided ecological balance, and the next year

---

* Also, could only ever be put forward by people unfamiliar with the Burning Man Festival in Nevada's Black Rock Desert.

saw a riot of grasshoppers that, uneaten in their youth by sparrows, grew to be an even larger menace. The Chinese, at least, got to stay with their fields and families; previous incursions of the state into the workforce have not ended so happily for the participants.

I'm referring to coerced work that occurs in the service of the state, labour brought into the city to facilitate some particular trade or many. One of the best-known bioarchaeological studies of slavery comes from the African Burial Ground in New York City. Michael Blakey and his team studied the remains of around 400 slaves who were interred in the burial ground in the seventeenth and eighteenth centuries AD. The project took care to consider the emotive context of the site, and to use whatever analytical tools they could deploy or develop to understand the experience that these people had. Bioarchaeological analyses revealed telltale evidence of enslaved African lives, confirming the historical record of the site as the 'Negro Burial Ground'. Skeletal morphology showed characteristics of African ancestry, while isotopic studies confirmed childhoods spent far, far away for many of the adults. Evidence of disease and particularly of growth disruption (the lines on the teeth that are my particular research interest) spoke to the wretched conditions that many of the slaves had lived through. Muscle attachments and other markers of activity testified to their hard lives, and palaeodemography showed the crushing reality of infant mortality among the malnourished and overworked souls buried there.

There are considerable variations in the nature of slavery throughout history and across cultures; attempting to summarise the work of slaves through the course of history is rather more than this chapter is going to accomplish. What we can draw out, however, is that the pattern of inequality that underlies the institution of slavery in whatever form it is practised will be reflected in the bones and teeth of those who suffer it. Debra Martin has looked at those fracture rates in the Ancestral Puebloan populations of the American Southwest and sees structural violence not just in the constant recurring injuries of a small subgroup of women indicative of domestic violence, but in the added evidence of poor health from signs of developmental stress and episodes of infectious disease. The presence of increased evidence of osteoarthritis and activity markers on bone singles these women out, in her view, as slaves or akin to such, suffering in death the further indignity of being denied grave goods like other burials.

One of the first pieces of evidence we have of coercive work comes from the dawn of the very first states, the urban cities networked together in the empires of the Sumerians, the Akkadians, the Egyptians and the like. The ancient Mesopotamian city-states, as we have discussed in Chapter 8, engaged in what we today would easily recognise as wars of conquest, and one of the main things they conquered was … people. They launched military campaigns to exert control over territory that they did not exploit in terms of agriculture or resource extraction, all in the hope of taking prisoner the ubiquitous enemy: the people of the mountain. Why, you might wonder, would a city want more people? Especially ones who probably didn't like you very much, judging by what you recorded *in stone* that you did to them, their families and their children? The answer, of course, is the same for the Sumerian Empire as it was for the British, the Portuguese, the Siamese, and the Maya. In a process that was played out by agricultural states from the third millennium BC to the second millennium AD, undesirable jobs required people who weren't quite people: slaves. The Sumerian word for slave was fairly clear on the subject: roughly translated, it just meant 'foreigner'.

The web of power and influence built up by the Sumerians was grounded in their vast productivity and populace, and sustained by trade up and down the rivers of Mesopotamia. This trade took many forms, but one of the key industries was borne on the back of our friend the sheep; at some point around 4,000 years ago, something like 50,000 people were employed in the Mesopotamian wool industry.* A significant

* Not to mention millions of sheep.

proportion of the workers in the industrial production of wool products – namely the female spinners in the urban factories – had their food subsidised by the state. While we have never successfully identified the skeletal remains of these inmates of the weaving industry of ancient Mesopotamia, we do have considerable evidence of the modern equivalent: the workhouse.

The workhouse was an institution developed by 'civilised' modern states to prevent the morally unacceptable situation of an idle poor.* They occur from at least the seventeenth century but become a staple of civic life in the eighteenth and particularly the nineteenth centuries in the British Commonwealth. As a form of poor relief, workhouses were considered the only ethical solution to the waves of mass unemployment following wars, famines and shifts in the productive economy. In phrases that would not be unfamiliar to followers of hard-right pundits today, the poor were frequently judged to be makers of their own misfortune, having committed the principal sin of being work-shy in an era where the very nature of work was changing. Workhouses offered accommodation and a limited amount of food in return for labour; towards the end of their existence in the nineteenth and early twentieth centuries, they were used more as statutory accommodation for those unable to work. Fans of Dickens (or at least musicals) will recognise the failings of the system from *Oliver Twist*, who impudently asked for more food than the workhouse rations he was allotted. Twist came out all right, but recent work in Ireland has shown that this was not usually the case.

The Irish Potato Famine of the mid-to-late 1840s was responsible for an incredible amount of human misery. Part of it† decamped for other shores, leaving for the possibility of America or even England. Those that could not or would not

---

* The modern equivalent of this derisive dismissal of the unemployed as 'skivers' has recently seen a massive popular resurgence.

† Including the illustrious Martin Hassett, authorial forebear.

leave suffered half a decade of food shortages, and the poor of course bore the brunt of it. This grim history has been revealed by excavations at the Kilkenny Workhouse of hundreds of bodies of those who died as inmates of the workhouse system. We discussed briefly the use of maize flour in Ireland in Chapter 5; the reliance on this low-cost food was to blame for the high rates of scurvy seen in the skeletons. The cost of malnutrition is written clearly in the slowing or even cessation of growth in the bones of the dead; there isn't even an uptick in the number of lines on teeth, something the researchers conclude is because nobody buried in Kilkenny survived long enough to form them. In the case of workhouses, the bioarchaeological evidence we have for the world of work in cities comes from the evidence of deprivation, and not from action.

One of the things that cities do well is to create a focus for the display of power to reinforce the symbolic rules and structures of their worlds. These are the urban corollaries of the Chinese animal wars, and across time and space states have stepped in with threats, bribes and entreaties to divert the labour of their people to these monolithic enterprises. One of the most pat examples of such a system ever to walk the long grass of history is the system of corvée labour at work in the Giza suburb of Cairo. An archaeologist in possession of a degree is an archaeologist in want of work, so in 2005 I hopped at the chance to go meddle in one of the most iconic locations ever to be dug: the Pyramids at Giza. Of course, we weren't actually digging the pyramids. This being modern archaeology, we were interested in social processes, not just pharaohs and pointy piles of stone. We were actually digging the workers' village just off to the side of the three main pyramids and the Sphinx, where the labourers who had done the work of actually *building* the pyramids had their village.

One of the benefits of archaeological fieldwork is embedding in another place and another culture. At Giza, this process had been bizarrely reversed. Because of Egypt's strict residency laws, and of course financial concerns, the vast army of modern Egyptian workmen on site, who moved the million tonnes[*] of sand shot through with three centuries' worth of camel-tour urine,[†] set up their own temporary camp on some unused ground. The workmen mostly came from far down south, which, confusingly, is 'Upper' Egypt. They were farmers on the ridiculously fertile shores of the Nile near Luxor, living in traditional mud-brick houses that are blessedly cool in summer, even if the sitting benches are far too high off the ground for someone my height. For generations, they had been spending the agricultural off-season earning some spare cash by hiring themselves out as labour to the ubiquitous archaeological digs both in Luxor and further down the Nile. Their rationale was obvious: the extra cash would pay for bride prices,[‡] farm equipment and new livestock. Our resident feminist Turkish trench supervisor Banu Aydinoglugil did try to raise a Marxist insurrection, but in the end the workers preferred to come up the river, do their work, get paid and go home.[§]

The archaeological workmen lived, worked and slept just hundreds of feet away from where the labourers who built Khufu's pyramid (the big one) had, some 5,000 years earlier. We could see the archaeological evidence of that occupation in the preponderance of high-quality meat for hard-working pyramid builders, in the presence of a barracks-like dormitory and an industrial kitchen for endless beery breads, and more broken standardised bread moulds than you could shake a theodolite at. The hand of the state is clear in both the size of the enterprise and in the carefully collected sealings that demonstrate the administration of the project on behalf of the state. The scale of the site is incredible; it was year 21 of

---

[*] I exaggerate.

[†] I do not in any way exaggerate.

[‡] In 2005/6 this was something like 15,000 lira for a girl educated to age 12; the price goes up per year of schooling.

[§] The Marxist still got invited to weddings in the ancient mud-brick villages the workers lived in down in Luxor, so presumably all was forgiven. Though they did make her sit on the men's side of the party.

excavation when I got there, and as far as I know they're still going. Five millennia ago, what were probably off-season agricultural workers were enticed with bread and meat; today, it's cash. At the modern workmen's camp, they must have been producing an archaeological record to rival Pharaoh's – bones of animals from parties, discarded *galabeya* and broken archaeological tools, and a mountain of 'Cleopatra' brand cigarette butts. The skeletal remains of the archaeological workers, however, will rest far away, home in their villages up and down the Nile; given the transitory nature of the work and with the fluid interpretations of occupation from the skeleton, we may never identify the skeletons of the people who built the pyramids. Sometimes, bioarchaeology has to ride shotgun to archaeology.

The type of work we do affects every part of our skeleton, just not always in ways we would immediately recognise. As I discovered following a recent conversation with a group of archaeologists with more or less academic interests in the archaeology of childhood, the way we work can even affect our children;* there is the suggestion that a transition from mobile life to more fixed farming led to the development of the cradleboard. Bioarchaeologist Siân Halcrow pointed me towards two examples of cradleboards from the Field Museum in Chicago. It's clear that one has footrests and one does not; this change relates to how the board was worn, and with the advent of agriculture the footrest may have been necessary because the infant would have been strapped to the back, but with the advent of more sedentary tasks like grinding things, the board could be laid flat. Depending on how they are used, cradleboards, intentionally or not, can actually reshape the malleable skull bones of the growing infant; a strapped-down skull might grow elongated and elegant compared to the boring noggins of the free-range baby.† The considerable

---

* Thanks to Siân in particular for flagging up changes in cradleboard technology.

† Adjust this statement for cultural preferences. If necessary refer to the diagram of dog skulls in Chapter 3 and think about which method of messing with your head is kinder.

cranial deformation present in archaeological remains of the indigenous inhabitants of the American Southwest is just one example of a practice observed elsewhere in the Americas and Africa.

The urban world offers a much wider range of jobs that an individual can do, as opposed to the somewhat monolithic employment of agricultural food production. Full-time specialism is a particular feature of dense populations; from the very first temple administrators of Mesopotamia to, say, the derivatives regulatory compliance technical reporting project managers of New York, the demands of agglomerative living require highly specific roles to keep the gears of the city moving. The opportunity to specialise, however, comes with its own dangers,* among which are the physical consequences of repetitive movements required by particular occupations. In cities we also see the creation of occupations with their own particular sets of dangers: workplace hazards and risks that would send a modern HR department into meltdown. The final problem with working in cities comes down, again, to the number of people involved. Cheap, easily replaced labour is rarely rewarded well.

---

*Just ask the nearest telegraph operator.

CHAPTER THIRTEEN

Panic …

This book has largely been about individual aspects of urban life: how changing ways of living have affected our distribution, density and diseases. For the final example of how city living has gotten into our very bones, we will look at the culmination of 15,000 years of creeping urbanisation: the birth of the modern city. Instead of going death-by-death through the global history of modern cities, this chapter will try to walk through the last few hundred years of urban history using the capital city of a small and standoffish island to see what there is to see: a history of population explosion, of global trade and local politics, of pollution and medicine, and all of the push–pull factors that have delivered the people of this city to the graves of paupers and of kings. With a wealth of history and an embarrassment of historians, archaeologists and archives, we will turn to the story of early modern London, and how the city described by Tacitus as a town somewhere below the rank of a proper Roman settlement became the city described by Dickens as a mountain heap of misery.

I love London, in the abstract way that only a child who grew up in a world where 'old' things date to the 1960s can. It's also the city I know best, having lived here for more than a decade. London was at the heart of my PhD research, and I'm forever grateful for the kindness of historian Vanessa Harding in letting me audit her MA courses on early modern history; much of what I know about the city is inflected by the court rolls, wills and other historical accounts she exposed me to. I'm in constant awe of historians; I'm not sure how you could come up with a coherent narrative about anything when the details of history are so delightfully revealing. Even in the process of composing this chapter, I got distracted by an account of a session of the Middlesex Magistrates' Court of one Mr Wotton, who in 1585 was running a school for pickpockets and cutpurses in the back of a pub near Billingsgate Market. He'd rigged a purse with bells all over and even created a graduated hierarchy of skills with titles for those boys who passed his tests; it's such a strangely detailed thing to know about someone dead 400 years – and it's distracting us from our story.

Despite their diverting character it's precisely these detailed, written and, most critically, preserved minutiae that allow us to reconstruct the history of the city so clearly. History, however, is a liar and a cheat when it suits; there is a reason that there are no glorious stone obelisks set about the temples of Mesopotamia carved with commemorative images of the time the Mesoptamian king was ignominiously defeated in battle. It's axiomatic that history is the past as told by the winning side. Only the dead never lie, but, as we've seen in the rest of this book, they certainly can mislead. I will attempt to pull together some of the endless bureaucratic details of the last few centuries, the little lives caught in their margins and the more human stories writ in the vast numbers of bones under London's streets, buildings, parks and homes. This should show us in a microcosm all that we need to know about life (and death) in the city: the how many, who, what, why, when and where that the issue of urban life throws out.

The foundations of London lie far beyond the early modern period we want to get at here, but merit some mention. London sits astride (it used to be aside) the mighty Thames, at a convenient point where the river is wide enough to get back out to sea but narrow enough to cross. Quite a larger number of early island inhabitants did cross it: there are footprints up in Happisburgh* on the northeast coast of Norfolk that have been dated to one of our hominin relatives more than 800,000 years ago. Britain itself is attached to the continent of Europe through the (now underwater) region of

* This is pronounced 'hays-burrow'. Place names in England exist for the sole purpose of tormenting non-local speakers. I mean, really – where did the p's go?

Doggerland,[*] which only appears when large parts of the planet's water have been taken up as glacial ice sheets. While it's hard to imagine today, my house sits on the edge of the vast glaciers of the last ice age that once stretched all the way into North London. A Britain covered in ice is actually a bit easier to contemplate than one covered in savanna, but just down the hall from my office in the Natural History Museum we have an entire cupboard full of fossil fauna dug out of Trafalgar Square in the 1950s – and it's full of lions and hippos that once roamed central London.

So it's fair to say that London has seen considerable change, even before modern *Homo sapiens* arrived. Geoffrey of Monmouth, who dedicated himself to writing a very detailed, very imaginary history of Britain, recounts the foundation of London as a story of the defeat of the giant Gogmagog by a misplaced Trojan. Archaeology has yet to uncover the remains of any giants, however, so the origins of the city are traditionally ascribed to the Romans at the start of their incursion into the British Isles. This is a bit unfair on the rather older evidence of occupation – or at least human presence – that archaeologists have turned up, particularly along the banks of the Thames, but the story of London as a *city* really does begin with the answer to Monty Python's question in the movie *Life of Brian*: 'What have the Romans ever done for us?' For the next 2,000 years, there were raids, razings, reconstructions and revolutions, leaving an extraordinary trail in the archaeological record. The city we want to speak about, however, is the city of the last few centuries: a city that has burst its banks and taken over both sides of the river, overflowing its walls to establish suburbs all the way from Westminster to the Tower of London. This is a city that was decimated by plague, going from possibly around 80,000 inhabitants at the start of the fourteenth century to little more than half that by the end.

The re-peopling of London after the Black Death was not a simple case of the production of more London babies by

[*] The etymology of which I'm sure is fascinating.

more London parents, as the eminent historian Paul Slack has demonstrated; he cites a variety of factors that kept the number of Londoners down after the catastrophic mortality of the plague. The factors he cites are interesting for how *modern* they sound to twentieth-century ears. Migration was a factor, with pilgrims, dissidents and others upping sticks and crossing seas, but the major brakes on the endogamous reproduction of the population are all attributes of urban life. There are the usual suspects in those diseases that flourish in dense, close-living urban conditions, but there are also social factors in reproducing urban life that resulted, not just from poor living conditions in London, but also from *better* living conditions.

The nature of work and the city, heavily influenced by patterns of migration, contributed to later ages at marriage than you would expect from the past; the average age at which men and women married and started families was in the mid-to-late 20s. From my research with Andrew Bevan into the demographic history of a rural community on a rather different island some thousands of miles and several cultural steps away (the goat-ridden paradise of Antikythera in Greece we visited in Chapter 3), I can see clear evidence that family formation was something that started much younger for the units of reproduction (*i.e.* women) in rural* agricultural communities. It's clear that the simple expectation that people in the past universally lived fast and died young is misplaced; then as now, circumstances dictate human choices on when and how to reproduce. In London, specialised employment in specific trades followed a system of apprenticeship that more or less took people off the marriage market for years at a time. Even for those who weren't overtly specialised into one of the established trades, the presence of waged work in the city was too great an opportunity to be missed, even if it meant delaying marriage and starting a family. The great financial draw of working in London brought in migrants from all around the country and even

* Very rural. Four-hours-by-boat-to-the-next-island rural.

beyond, but they came either as young men who would need seven years to become their own master in the traditional guild system of tradesmen, or as domestic servants, particularly women in the latter part of the period, putting off marriage and starting a family for the surety of hard cash.

All this cash is one of the key features of the modern city. Wages – earnable, transferrable wages – operate in a system of exchange that recognises no difference between luxuries and necessities – they all have a price tag. Abandoning this cash economy and living off the land in London is not an option.* The urban economy is fully moneterised, made of credit, capital and, most of all, cash and the ability to earn, raise or steal the same.† Repeating a refrain that was still current at least the last time I spoke to my grandparents, contemporaries of the early modern period deplored the 'spirit of madness running abroad, and possessing men against marrying'. Even more pertinent to the modern participant in the cash economy, the seventeenth century describes the plight of the potential parent 'afraid they shall not be able to maintain the children they shall beget'. Anyone who has ever looked at the price of a pram will understand the dilemma.‡ The engines of global capitalism may have been running on timber, coal or steam, but the fug of money problems they emitted into the urban world seem not so very different to those of today.

There is of course a final factor in holding population numbers down in the early modern period, and that is death.

---

* Which is not to say that people haven't tried. I recently had lunch in a converted skip on the construction site around King's Cross station right next to the plants that had been denuded for my salad.

† Again, there is opportunity to get lost in the assize (court) records here; part of their charm might derive from the fact that at least some of the sessions were held in the Mermaid Tavern, infamous drinking den of the Elizabethan era, which used to be somewhere near modern Bread Street and Cannon Street. Justice in the pub sounds more fun in general, though it did seem to end in death very frequently, so perhaps not for all.

‡ Seriously. You could buy a car for less. I certainly have.

I think the fact that we're not surrounded by 200,000-year-old hunter-gatherers is reasonable proof that death is not the exclusive province of the city. But the rather fantastic thing is that sometime in the 1500s history begins to record what previously only the bones themselves had. Bureaucratic compilations of births, marriages and deaths were recorded by an officious and jealous Church slowly supplanted by an interventionist state. And this habit of keeping vital registers of the population was further extended to keep an eye on denizens even as they departed this world. These are the Bills of Mortality, and they're the reason we know so much about mortality in early modern London.

The rather macabre collection of statistics embodied in the Bills has been mentioned briefly before, but they are remarkable documents. The point of them was to record plague deaths and act as a sort of early warning system for the powers that be if an epidemic approached; they were used in many European cities in the early modern period. Plague was still very much a threat at this time, cycling through Europe and causing catastrophic mortality until the last major outbreak in Marseille in the 1720s (another big, connected city at the height of its commercial clout). However, as the Bills themselves record, plague was certainly not the only thing to die of in the early modern city. The best-known collection of Bills records the plague year of 1665, the one that Samuel Pepys found so very pleasant. The conditions recorded require some flexibility with the facts of modern medicine – who dies of 'horseshoe head'? – but they reflect contemporary attitudes towards disease. For just the week of 7 March 1665, various causes of death are recorded, such as: 'aged', 'cancer' and 'childbed', though 'consumption' is there as well as the 'King's Evil' (aka scrofula, see Chapter 9). 'Mother' is also down as causing a death, and there is someone 'mouldfallen'. 'Rising of the lights' appears to have taken off nine souls, and scurvy and rickets also do their worst. The 'French pox' is indicted in four deaths, in contrast to the small one that carried off 16. Sometimes the Bills give a very clear cause of death – 'killed by a cart by St Martin-in-the-Fields' – while others are touchingly esoteric: someone in London, in the week of 7 March 1665, died of 'grief'.

They also reflect the medical knowledge of those who recorded them, a group so disparaged by the male medical establishment that we will have to take a moment to wade through the disdain to try to get at how accurate these causes of death might be. In the sixteenth century, the tradition of parish churches employing 'searchers of the dead' was instituted. These searchers were, quite literally, employed to search for the dead: to visit the same houses as Death and to ascertain what form it had taken. They were almost always elderly, indigent and female, which meant that they had all of the criteria to take on a job that, on the surface, sounds relatively undesirable. As poor women, they would have become dependents of the parish and been obliged to take what employment there was. No doubt having survived to old age, there was a good likelihood that they would survive future contagion, and even if not, there were plenty of other little old ladies around.* The very early epidemiologist John Graunt, who pulled together many of the Bills of Mortality in order to make one of the first use of statistics in the understanding of disease, described them as 'antient matrons, sworn to their office', but later historians were even less kind. Popular perception seems to have been that the searchers were uneducated illiterates without the wit to put together cause of death. An anonymous correspondent to *The Gentleman's Magazine* in 1799 reports that in at least two parishes of his knowledge, 'the searchers cannot write; the mistakes they make are numberless in the spelling of Christian and surnames, for, they trust to memory till they get home; then, child or neighbour writes what they suppose it to be'.

Still, you wouldn't want to be on the bad side of the searchers. Cause of death matters to the living relatives and associates of the deceased, not least in times of plague. If the

---

* This would be the extreme version of 'workfare': take anyone living off the parish and see if they survive visiting plague houses.

searchers declared a plague death, quarantine and restrictions would fall on the household; the economic and social consequences were harsh. But how much reliable information can be got from causes of death that seem to be a collection of euphemisms for diseases that don't even really exist? More than might be thought on initial reading. I've already mentioned in Chapter 10 the peculiar case of 'rising of the lights' being much more comprehensible when you realise that 'lights' are 'lungs'.* The searchers may have reported in the vernacular, but almost all diseases were vernacular in the early modern world. Take for instance our baffling case of 'mouldfallen'. This is probably linked with diseases, recorded on the same line elsewhere in the Bills, such as 'mouldhead' or 'horseshoe head': forms of hydrocephaly. Hydrocephaly – water on the brain – causes a type of cranial swelling that would have been noticeable to even a lay person, and hydrocephaly in London was more than likely to have resulted from infection of the central nervous system with our old friend *Mycobacterium tuberculosis*. Death by 'spleen', however, shows that the opposite can also occur; what sounds like a practical appraisal of an anatomically limited problem is in fact likely to indicate a choleric distemper (more antique words implying medical diagnoses) – in other words, depression.

Without germ theory, Londoners lived in a pall of disease caused by all sorts of things. Worms in teeth were at the root of tooth decay, an idea that has been with us since the Babylonians. But above all, the environment was responsible for all manner of evil. The idea of humours had not gone out of medical fashion, and the climate, the temperature, the disposition of land and the fabric of buildings were all blamed for causing different ailments. The streets were thoroughfares of filth, and it must say something for the early modern conception of hygiene that the water running down roadside ditches was considered for use as both toilet and watering

---

* So no, punching someone's lights out does not mean what you think it does.

hole – something we actually know obliquely from two unfortunates who died in the same open sewer (or urban stream, if you like), but at the very different purposes of making and retrieving water. The air itself was not to be trusted, being full of vapours and, worst of all, the dreaded miasma.

But how bad was this brave new urban world? How dirty was it, compared to all of that skulking about in caves for all those hundreds of thousands of years* before cities? Let us start with the filth on the ground before raising our eyes to the skies. Despite most of us holding a general perception of the past as a more bucolic place, full of greenery, wild animals and fluffy clouds, we seem to only ever imagine the cities of the past as potentially dank and polluted places. Picture the scene changes of any knights-in-armour type film and you'll see what I mean – rolling hills transition through a portcullis to wretched filth within the city. There may even be ragged lepers for effect. Really, only the insightful work of Monty Python has dared contest this seemingly inbuilt urban prejudice. Their 'mud farmers' scene in the film *Monty Python and the Holy Grail* manages to get right down into the muck of rural life while taking pot shots at the social processes that put people there.† As we discussed in Chapter 3, there are pollution risks associated with close-living livestock that can lead to increased disease transmission; these might be expected in contexts where people cohabit with their goats, but those contexts may or may not be urban. Agro-pastoralists, professional herders who also do a bit of farming, may have constant close contact with animals but are hardly urban dwellers. And in the space-conscious city,

---

* There were non-cave moments in between.

† This scene also works as a critique of feudalism and/or mysticism; in the words of the 'male' character: 'Strange women lying in ponds distributing swords is no basis for a system of government! Supreme executive power derives from a mandate from the masses, not from some farcical aquatic ceremony!'

there just isn't the room to keep that herd of goats with you at all times.

What cities do manage to do with the evocatively pastoral pollution of the animal world is build it up into terrifying quantities. Cities are aggregators – of people, things and animals; and until the advent of modern refrigerated transport, if you wanted to eat an animal, the animal had to come to you. Awkwardly, as cities got larger, if you wanted to get the animal, you increasingly had to get other animals to take you there. Animals have been present in the urban world since before it was properly urban; we can think of those goats and sheep at Aşıklı Höyük huddled in close some 10,000 years ago. The intensification of animal stockherding that really took off in the last 3,000 or 4,000 years in some places was never abandoned, leading to entire cultures of mobile pastoralists.

Alternately, we see a long tradition of herders so integrated with a sedentary lifestyle that we don't view our pastoralist pasts as being separate from the story. If you ever find yourself at a loss for something to do in the French Pyrenees for a few days, I can highly recommend reading the strange story of Montaillou and seeing what an odd but not uncommon blend of nomadic and settled life looks like in practice. Montaillou was (and is, you can visit) a very small town on the fringes of France, about 40 kilometres (25 miles) from Andorra and just over the ridge from Spain. As a result of a scheming village priest, some unfortunate love affairs and an *incredibly* meticulous Inquisitor, the testimony of more or less the entire village was taken during the tail-end of the Cathar Inquisition in the years between AD 1318 and 1325. One Pierre Maury recounts his life as just such an agro-pastoralist, somewhat tainted by his association with the heretical Gnostic beliefs of the Cathars. His response comes to us in incredible detail thanks to the lead investigator (and future Pope of Avignon) Jacques Fournier's idiosyncratic method of going about quashing heresy by basically just asking an insane number of questions and writing down all the answers, a fairly unique take on Inquisition in a movement with a reputation for inventive tortures. Pierre's life as an illiterate shepherd 700 years ago is full of unexpected drama,* but it's also full of mobility and the breadth of experience is striking. He wanders through all the Occitan lands, which today cover at least three countries; he lives for months with Muslim workmates, visits cities big and small and quits his jobs whenever he feels like it, everywhere taking sheep to and from hills to shearing. Life in the city seems stultifyingly stationary compared to his experience, whether you were the human or the sheep.

Pierre's story touches on an aspect of urban life that appears in the Neolithic but then slowly mutates as populations become denser and work more specialised: animals. Despite many urban dwellers maintaining a small personal menagerie, contact with *living* animals was probably more limited than in rural, pastoral contexts, with the potential exception of experimental early villages just developing the domestication process, like those at Aşıklı Höyük. Nonetheless, the big urban cities of the past still drew in meat to market. For millennia, drovers moved their flocks through cities, letting the animals do the work of marching on to their deaths and giving city dwellers the opportunity to eat. There are specific routes that were used by stockbreeders going to market still riddling the modern world; even cosmopolitan London has an archaic statute on the books granting the Worshipful Company of Woolmen, the ancient guild of sheep keepers, shearers and wool sellers, to drive their sheep across London Bridge, a right they rather surreally exercise once a year, much to the bemusement of tourists.†

Today in many parts of the world this is still an important practice; I will not soon forget spending Eid al-Adha amid the crumbling colonial facades of half a dozen empires in the undeniably Mediterranean but defiantly Egyptian port city of

---

* There are paternity issues and complaints about getting girls to come up to the pastures for the night.

† It is rather fantastic: https://sheepdrive.london.

Alexandria. Eid al-Adha is the Islamic Feast of Sacrifice, commemorating the biblical story of Abraham, who, asked by God to sacrifice his only son, is rewarded for his homicidal piety with a miraculous kid–ram switcheroo. This is generally celebrated by acquiring or choosing a sheep in the preceding week, moving it in, getting the family and especially small children to love and pet it, and then slaughtering it in front of everyone.* In Alexandria, like in many modern cities, people live in apartment towers and blocks, and sheep keeping is slightly out of reach,† so enterprising herders bring their sheep to town. For convenience, they will also do the slaughtering for you, which is very considerate until you realise that bleeding out an average of about 30 sheep on every third street corner quite literally makes the streets run red with blood.‡

The thoroughfare of animals in cities might more habitually leave behind animal excreta, but butchery too makes for a good source of animal pollution. By far and away the biggest contributor to animal pollution in the city, however, is the excreta of our transport animals. At one point in 1907 the American city of Milwaukee was producing 120 *tonnes* of horse manure scraped off its streets *every day*. Many enterprising cities invited farmers to come and collect the deposited wealth of urban animals; a hundred years earlier, New York City had sold 29,000 dollars of horse manure to farmers. London too had an active 'nightsoil' industry, supplying outlying agricultural areas and the tanneries on the eastern and southern fringes of the early modern city.

Casting our eyes away from the filth under our feet, we have at some point got to consider the air. Certainly, Londoners

---

* We could have an entire subchapter here on traumatic experiences of Eid al-Adha (or Kurban Bayram in Turkish), but I think we're sufficiently off-topic for now.

† Though Cairo seems to be home to a particular breed of solely roof-dwelling caprids.

‡ The only time in my life I have regretted wearing Converse. The white band at the bottom turned bright pink.

always did. They were convinced that the air was the source of every contagion, every cough and fever. It was an established medical fact in the early modern period that disease was caused by 'miasma', a pestilence spread through the air, and that miasmas hovered in fens, marshes and other low wet ground. London being full of such places, it was little wonder that disease stalked its citizens. The early modern belief that miasma in the air had to be kept out may have been connected to the presence of a terrible fever that presided over the then-marshlands of the eastern coast of England. It affected places like Essex so badly that Daniel Defoe, another keen diarist of the seventeenth century, was told by a group of 'marsh men' with straight faces of varying degrees of believability that anyone born outside the marshes was likely to die shortly after moving in. This led to the happy situation that a marsh man might expect to have five or six wives over the course of things if they brought wives in from 'the uplands'. In fact, the English had their very own form of the French *aguë*, which we know today as malaria. The identification of the source of the disease – the word miasma means, literally, 'bad air' – was only half wrong; it's generally agreed that plasmodium bacteria carried by mosquitoes were at one point endemic to these shores. Destroying the heart of the fens may have wreaked havoc on the ecosystem, but it did have the notable benefit of ridding England of a pestilent home for mosquitoes in the malarial swamplands just to the north of London (and depriving Essex men of their opportunity for new wives). Of course, at the time, it was the miasma that was considered vanquished, because everyone knew diseases lived only in the air.

It turns out that it might not have only been disease wafting around the streets of London. All those people, all those houses, all that industry, created an extraordinary amount of air pollution. People think of the 'pea soupers', the great fogs of particulate matter that used to turn day into night in the coal-burning capital of the mid-twentieth century, but smoky domestic fires of wood, peat and animal dung have been with us from the very beginning. In Chapter 2 we saw what happened if you try to occupy your new-build Neolithic dugout house without properly checking the ventilation: hordes of archaeologists choking and running for the exit. Actually, smoke inhalation seems to be one of the primary dangers of experimental archaeology; experiments at Çatalhöyük and Aşıklı Höyük in reconstructing the hearths of their ceiling-entry dwellings seem to have suggested that people and fires could not very easily coexist indoors.* It's not clear that Neolithic housing was all that much of an improvement on living in caves, air-quality-wise. While it's difficult to reconstruct air pollution factors from tens of thousands of years ago, especially if you only have an archaeologist's floor plan to work from, one of the interesting suggestions made by bioarchaeologists has been that we might be able to pick out signs of aggravated snuffling due to poor air quality from subtle changes in bones. The nose and mouth are where the human body takes in air,† and these are lined with delicate tissues designed to protect the even-more-delicate tissues further down the line by responding floridly to threats both real‡ and perceived.§

Constant irritation and inflammation can cause sinusitis – infection of the sinuses – and sufficiently irritate the flimsy bones that surround the nasal passage and sinuses so that they develop porosity on their surfaces, just as in any other inflammatory response. Noting shifts in levels of sniffles through time by collecting bioarchaeological evidence of inflammatory response on these bones has been argued to be one of the few ways we can reconstruct the quality (or lack thereof) of our housing, from cave to condo. The pioneering physical anthropologist Calvin Wells extended his investigation of palaeopathology to the sinuses of the upper jaw, sticking an endoscope up the noses of medieval British skeletons.

---

* Sadly I never got to see many of the experimental reconstructions of either Aşıklı Höyük or Çatalhöyük, though I was treated to a live re-enactment of the highlight reel on one of the Çatal party nights.

† Unless something has gone horribly wrong.

‡ Microbes, pathogens.

§ Cats.

Subsequent work by a group of British bioarchaeologists including Mary Lewis, Charlotte Roberts and Keith Manchester has produced a synthetic picture of air pollution in the British Isles over time. Roberts has compared the experience of air pollution across several widely spread groups stretching from North American hunter-gatherers to nineteenth-century London. She found that outside of the city, women had higher levels of sinusitis than men, but that urban living equalised each sex's exposure to irritants, and also led to more exposure overall. If the home, urban or not, can be a source of pollution, then the elevated levels of sinusitis in non-urban females compared to males that we see in both archaeological and modern evidence might directly relate to the division of labour between the sexes. This pattern kicks in well before the Industrial Revolution, suggesting that the background levels of urban air pollution are of greater antiquity than those blights of modern air quality we point the finger at today: industrial manufacturing, power generation and transit. They stretch to the first urban agglomerations, suggesting that something about urban life – potentially the concentration of air pollution in homes or in city streets – is responsible for declining air quality. Certainly by the early modern period there were a number of industrial practices at work in the city that might have added their own particular flavours to the pall of smoke over the capital, including tanning, brewing and lime burning.

The pattern does not seem to be overly strong, however, suggesting that there might be a number of confounding factors between actual air pollution and the bony response bioarchaeologists can see. A study of pre-state agro-pastoralists and medieval Indian skeletons suggests that males were more likely to have sinusitis, but the pathology occurs so infrequently it's hard to draw a conclusion as to what that might mean; the authors suggest that dental health and other infections have as much to do with sinusitis as air pollution. It's a pity that larger samples were not available from the agro-pastoralists, as I would suspect that a life spent that close to animals would be tough on the schnozz in many ways, particularly reliance on animal dung for fires that then release industrial quantities of pollution into the home.

There is one further element in the story of London pollution that we have not yet covered: water. It's not that it's entirely unremarked on in contemporary times, but it seems to have been subject to a much lower threshold of purity than might be expected – recall the bodies found in the ditched streams running along main thoroughfares. The water of London was, simply, filthy. It's a wet city at the best of times, surmounting a larger river and burying several smaller ones under the foundations and basements of the city. These were not the only channels coursing under the early modern city, however. The Romans had brought plumbing with them, giving London nearly two millennia to develop a functioning hydrological system, including options for temperature control,* and riddling the city's substrata with sewers. Alas, by the medieval period, the Roman sewers had gone the way of the Roman temples and markets: lost underground, unseen and unsuspected. By the early modern period, access to water for the vast majority of the population was through wells and public pumps, or straight from the river itself. This would have been fine, of course, if only the majority of human waste produced in the city of London did not drain right back into these sources.

This brings us to the final example of death in the city that we will cover: a story that starts with death and destruction and ends with, well, destruction. But with this destruction nevertheless comes some signs of hope for our urban world. It starts in India. The bacteria *Vibrio cholerae*, which seems to have been endemic to the subcontinent since the beginnings of medical writing, burst onto the global scene in the early nineteenth century. The infection can cause no symptoms or

---

* Something that has, bafflingly, never returned. Why must UK taps be limited to one hot, one cold? Can there be no middle ground of pleasantly warm water, as there is in other countries? This is a key source of vexation for the expat, alongside more tolerable cultural idiosyncrasies like the language and food.

many. Cholera is characterised by an extreme diarrhoeal response; obviously this is in its favour, as the main mechanism of spread is in water contaminated by human faeces. The disease is particularly dangerous in children and other less immunocompetent people, and before the advent of rehydration solution, the loss of fluid could prove fatal. As with our other epidemic diseases, it's the cities that sustain and circulate the disease, and cholera arrived in London in 1832. This being the nearly modern world, its arrival was announced in the newspapers: there had been cholera in Moscow for a year, and there were reports of earlier outbreaks in India.

The response of the good people of London was the same as in every other epidemic terror: a great deal of public verbiage was expended on the threat, inadequate amounts of money were promised to the cause and everyone avoided everyone else like, well, the plague. On the occasion of the 1832 outbreak in London, one noble gentleman refused to come out at all, and made deliverymen chuck everything to him from the street. Some 3,000 people died in the outbreak, and while the consequences for them were of course quite permanent, there were no major shifts in the political or environmental response to the disease. After all, everyone knew that diseases came from the air. There were plenty of diseases that caused diarrhoea, many of which could even kill you. Typhoid, for instance, was another disease out of the subcontinent that could kill you in a similar way, but with the occasional addition of a red-spotted rash on the chest. Typhoid, however, was a known threat of considerable antiquity – recent (though contested) ancient DNA work has suggested that it may even have been responsible for the plague of Athens in the fifth century BC. The one thing that everybody knew in the 1830s, from the uneducated coal pickers to the august and learned physicians, was that diseases came from foul smells and noxious vapours; and these were the sources of any scourge. Prevention of disease could be achieved by cleaning the filth from the streets, or setting fires on street corners to counteract epidemic airs. This was an attitude almost totally unchanged from the days of the Black Death, where it was thought that handfuls of herbs might keep away that most dreadful of visitations. The alternative theory was that disease came, if not the air, then from people themselves; the proponents of the 'contagion' theory saw disease as spread by touch, like flu. Both theories were voraciously expounded in the medical and quack-medical arenas, and as the miasmists and the contagionists set to in the press and in the lecture theatres, people continued to die.

The periodic outbreaks of cholera in London were watched over by a promising young medic by the name of John Snow. A meticulous sort of man, he pored over the details of each outbreak, slowly compiling evidence to support a theory that had taken up residence. Snow was convinced that cholera spread not by air, not contact, but by *water.* He mapped the deaths of an 1848 outbreak and found there were more deaths in those neighbourhoods that were supplied with water from the most polluted part of the Thames. In 1849 he published this revolutionary idea, to which the general response of the learned community of miasmists and contagionists was that he knew nothing. At the very least, he needed to prove beyond a doubt that water carried the disease.

It would be five years, but Snow would get his chance. In 1854 the neighbourhood around Broad Street (which is now Broadwick Street) formed the epicentre for a cholera outbreak that was remarkable for its virulence. In just one night, 70 people perished in between a handful of streets. Entire families died, shuttered into their houses, not to be found until the smell gave them away. Snow, who lived relatively near the outbreak, trotted around collecting samples of water from the community pumps in the area. He also turned to his maps. He carefully plotted 578 cholera deaths on a map of the neighbourhood, tallying blocky lines at the doors of each of the houses to represent the dead within.

He noticed two things that were enough to sway him and the tide of history. First, the destitute inhabitants of the nearby workhouse were largely unaffected by the outbreak. Since it's usually the poor who suffer most from the burden of infectious disease, this was remarkable. However, the workhouse had its own well, and for once it was a blessing to be interned there while death stalked the more prosperous families all around. The other clue was the robust health of the brewery workers who lived towards the east edge of the radius of deaths. They failed to sicken because they took their drink in the form of beer, made locally at the brewery, and most importantly, boiled past the point of *Vibrio cholerae* survival in the process. Snow had his evidence, and went to the Board of Guardians of St James's Parish with it. The next day, they took the handle off the pump at Broad Street, and the epidemic was quelled.

This was a triumph. A triumph for science, and of course for Snow, but also in the role of the city in setting and demanding standards of hygiene and behaviour. In the aftermath of Snow's proof, the entirety of London's water supply was thrown into doubt. The sewerage and piping systems were rebuilt to standards that would have impressed even the Romans. I live in a house built in 1898. The pipes that run out of my sink and my bath and my toilet all run into a beautiful Victorian tile-lined sewer, and thence onwards to the great sewer system the city was forced to build. And that is the crux of the thing. We have looked long and hard at the changes cities have made in us, but in the last hundred years, we have changed cities. The open sewers have become waste-treatment plants. The carts of muck that roamed London's nighttime streets are replaced by two cheerful dudes in maroon uniforms who come around on a terrifying truck every Tuesday and tell me off if I put tin cans into the recycling. This house was built for the health of the Victorian worker, and despite the lack of tepid water and the constant threat of damp, it's surrounded by green spaces, connected by footpaths and public transport that doesn't poo directly on the street, and has yet to give me consumption, typhoid, cholera plague or smallpox.

I am a little worried about the sinusitis, though.

CONCLUSION

Karma Police

At some point, this book has to end. It would be nice if we could conclude with some definitive answers to the questions posed at the beginning. Why did we go from small groups roaming the landscape to giant ones packed into apartment blocks and overcrowded commuter trains? Was it even a good idea? What evidence do we have? Is it enough?

I've suggested that history is a liar and bioarchaeology a more devoted servant of the truth, but the reality is that we can only investigate what we can find. The trouble comes when we do not know what it is that we're missing in the body of evidence of all of our thousands of years of bones. Thinking back to the burial of the suicide by gunshot in the crypt of St Bride's in Chapter 7, we can see explicit rules of society clearly and definitely recorded by history being flouted by friends, family and community undergoing the real experience of a young man's suicide. Just a hundred and a bit years earlier, however, the sad case of Amy Stokes is a lesson in the opposite direction. The unfortunate woman hanged herself in early September 1590 and, found guilty of such a mortal sin by an inquest, was taken out to a crossroads and buried at night with a stake through her heart. We must be very, very clever indeed if we think to account for the amount of variation possible in the afterlives of the dead within societies, given all the differences that existed in the lives that have occupied our great big planet.

The late Peter Ucko's seminal 1969 work on the many different ways to dispose of the dead practised in the world, past and present, doesn't even touch much on archaeology, but has vast implications for how we understand lives in the past. In it, he describes ways of getting rid of the body that even a decade of CSI spinoffs haven't come up with, all of which are perfectly normal in their societies of origin and all of which leave larger or smaller gaps in the skeletal record. Exposing the dead on arid hillsides for vultures, caching them on platforms for leopards or taking them out of the grave every year for a bit of a dance: these are just some of the ways humans live with their dead. Without careful thought and strategic sampling, we will always miss part of the story.

In looking for the answers to this book's big questions, the first thing to understand is that we really don't know it all yet. This applies particularly to the evidence of our earlier hominin escapades, but most of prehistory is a selection of anecdotal finds; we don't leave obvious enough scars in the earth for archaeologists to find easily until we start to build permanent settlements.* Of course, we certainly do find remnants of our life before settling down, but we may never have as strong a grip on where and how many people lived lightly on the land as those who burrow into it and stay there. Imposing settlement mounds are easier to spot than little scatters of stone tool debris, and the history of archaeology has not always favoured the little guys. One of the things to be said about Childe's Revolutions – Neolithic and Urban – is that they were theorised on the backs of macro-scale artefacts: pots, buildings, things you can see with the naked eye. Now that archaeologists, myself included, spend so much time reconstructing the micro-artefacts of the past (pollen, seeds, cellular growth structures, even DNA), we find we have to rewrite the 'big picture' of how we first moved to no longer moving.

In chapters where we focused on the Neolithic, the revolutionary consequences of this new way of life were fairly clear. Childhood diseases, dental decay, even a loss of height all accompanied our first experiments at living agriculturally. Depending on what set of evidence you focus on, you could make the argument, and many have, that villages are the

* To everyone who has ever worked on an isolated microlith scatter or at the bottom of a terrifyingly narrow cave system: I salute you.

deadliest invention we have ever come up with. The category of evidence I know best is teeth, and we saw in the first few chapters that teeth especially record the struggle to survive in those early Anatolian villages. Those poor Aşıklı children were hit by some sort of disease, malnutrition or combination thereof more often than we think their foraging forebears were. And the children at Çatalhöyük, who were even more firmly embedded into the Neolithic way of life, may have suffered even more.

And yet, and yet. As our French dentist with the insatiable curiosity about syphilitic prostitutes says, the teeth of the children hold the secrets of the parents. It's true in two ways for our Anatolian skeletons. The systematic timing of the lines on their teeth suggests that they were getting sick at regular intervals – possibly when siblings were born. The other thing to consider is, of course, that lines on teeth might mean sickness – but they mean sickness *survived.* Sicklier, scrawnier, but alive to produce another generation, and with more siblings besides. Cycling up by several orders of magnitude to a full-scale map of the Near East and Europe, we see the success of the Neolithic writ large in the numbers of people it produces, and in the successful spread of the genes it carries with it. We don't know enough yet about Neolithic population scales and genetic flow in other parts of the world, but the data is coming, and we will watch it carefully to see if the faint suggestion that some Neolithics in some parts of the world were easier on our bodies than others really holds.

But the Neolithic is, of course, just a warm-up to the business we have at hand: our increasingly urban lives. I have skated around defining a city in this book because, frankly, I don't see that it helps to declare a particular population size threshold or rattle off a list of key traits.* What we need to know is which *features* of settled – then increasingly urban – life are the ones that change us, kill us and leave their traces in our bones.

* Also, it's very, very difficult and escalates to esoteric social concepts very quickly.

This book really hinges on Chapter 5, the one about inequality, because that's at the root of every subsequent chapter and is the defining feature of city life when it comes to our bodies and our health. Inequality is the result of specialised roles in society that get codified into differential access to resources. Specialisation isn't limited to cities per se – people apprenticing and then working full-time as shamans or craft specialists have also been important in more mobile communities.* But think for a bit about those new walls that go up, turning public space into private and shamans into temple priests. It takes a *city* to support a temple, which doesn't just have one 'big man' to go with it but an entire class of bureaucrats and adjutants and priests and hangers-on and the people that fetch coffee for them in the morning. You can bet your bottom clay tablet that the head priest at the temple of Inanna eats better than the poor sod who has to fix their half-caff almond milk latte.† That difference is real, even if I'm being facetious about it, and it manifests physically, in our very bones. Cities foster social roles that are more stratified and more profoundly separated from the means of production (where and by whom all the food and other stuff get made). Limit people's access to resources and you in turn limit their health, undermining their natural immune response via malnutrition.

Turning to the violence we have discussed, against individuals or against entire peoples, this occurs in all of these types of settlements and cities, across time and across the globe. The head-smacking and the genocidal campaigns actually predate urban life by some margin, with clear evidence emerging in the archaeological record that people have a long history of being, well, terrible people. But, having

---

* Much of the pottery we see in the prehistoric world that looks so similar we refer to it as a 'type' may have in fact occasionally been made by the same people, or carried by the same people – all these itinerant pots going around messing up our concepts of territory, trade and mobility in the deep past.

† Always tip your barista.

gone through the old arguments, it's hard to disagree with Keeley in Chapter 8, who saw us as red in tooth and claw but, crucially, *getting better about it*. I'm not optimistic that people will ever stop smashing each other – with fists, with pint glasses, with baseball bats. Nor do I look around the shattered lives and pockmarked landscape of places I've lived and worked in and see an end to war. No one could look at the fate of the Yazidi women captured by ISIS in their rampage through the former Mesopotamia and argue that people are not dying as horribly as they did 5,000 years ago. I would like to be able to say that we can follow Keeley's thesis that war is no longer as fatal as it once was to a happy conclusion: that states have a pacific effect. This would mean accepting that the numbers and types of people killed have narrowed; that war is between great nation-state combatants and not some sort of community genocide where you smash the legs in just for good measure. But there are massacres in this recent world too: in Yemen, in Syria, in Sudan and other places where the state has failed but the villages, cities and people remain. And when we think of mass causalities, do we forget that it wasn't just fighting-age men in Nagasaki and Hiroshima? If this is the *pax* of a global nation-state system, it's too wretched to take any joy in.

If you think about the types of structural violence, the kind that comes with a power differential, the vast majority are now on the way out. Not fast enough, and in some places the numbers have not moved an inch, but the very fact that we have a medical and *legal* framework for discussing child abuse is a remarkable change from the hundred years previous. There is an increasing trend away from corporal punishment in Europe; it has been banned in schools and even homes in some countries. The UN Convention on the Rights of the Child is, yes, an airy bit of UN-speak, but it enshrines principles we might actually all someday adhere to. Domestic violence aimed at women is another area where the crushingly morbid statistics hide the flickering lights of hope. There are cultural changes happening at a glacial pace, but again, legislation is slowly wending its way into ever more countries.

The funding for organisations like IMECE, the local domestic violence charity that helps members of the Turkish-speaking community near where I live, has been slashed and burned in the name of austerity politics, but I still see their employees going to work every day and giving their free time at night to women's rights issues around the globe. Campaigns against marital rape, child marriage and all forms of domestic abuse have entered the public awareness, not just in cushy Western Europe and North America, but on a global scale.

One in three women experiencing domestic violence in her lifetime is still abhorrent. But there are reasons for optimism; I for instance take great joy in the work of Mona Eltahawy, who manages to be a feminist just fine, communicating her message in societies usually dismissed as being unsalvageably misogynistic. I think there is actually a slightly paradoxical argument for the main agent of structural violence – the state – being in more recent incarnations the agent of *decreasing* structural violence. The abandonment of the death penalty, and the prosecution of extra-state, extra-judicial killings like those of so-called witches, suggest a pattern where state-sponsored structural violence is decreasing, not increasing. Again, not fast enough for the people caught with the rope (or the tire) around their necks. And there are extremely worrying trends towards the use of drones or special forces to carry out killings that would be illegal in the state that ordered them. But where once capital punishment was the norm, it's now a last resort, or even, in the case of many countries, the subject of an outright ban. It's hard to imagine any modern nation-state returning to the days of plucking the bloody, still-beating hearts out of prisoners of war, so perhaps we can count that as progress.

Cities also play many roles in a larger political landscape. They might be regional hubs, drawing in the loyalty and resources of a territory that lacks clear borders or edges based on geographic or cultural convenience. Cities might exert very little power beyond their own domain. Or cities might be parts of a larger whole. Think of Caffa, the Silk Road city that sent plague back to Italy, which had all the hallmarks of urbanism (dense population, walls and a bureaucracy), but that just happened to be run by people from hundreds of miles away. You could have dropped Caffa on the moon and it still would have functioned, as long as there were trade routes to sustain it.* Cities might also be cogs in the regional machine, instruments of a larger power. Alternately, they might stand alone, pinned into limited territory claimed by rivals of similar size and strength.

Cities draw in population through some arcane formula fought over by social scientists for hundreds of years. Part of this is native internal growth, a phenomenon we have discussed mostly in terms of the Neolithic Demographic Transition: an increase in agricultural calories and the reduced requirements of staying put spur birth rates upwards. However, growing past the scale of villages and into the heady numbers of the more properly urbanised world, cities incubate not just people but all the things that kill people: inequality, disease, other people. These are the three things needed to unleash the catastrophic mortality of epidemic diseases and the horrific physical consequences of some of our more disfiguring endemic ones. We can actually see the role that our cities play in controlling which disease we get in two ways. In the first, we see the diseases of density: diseases that need a big enough reservoir so that they can burble away in the background, slowly making your fingers rot off. For the second function, we see the diseases of enhanced urban connection: take all of these lovely endemic problems that we've been carefully cultivating and fling them over the parapets at whomever we meet. The decimation of the Americas was set in motion by the globalisation of the urban world and by cities in contact with other cities, constructing rapacious nation-states on the back of resource acquisition and competition. It's not hard to imagine that our globalised network of disease could come back to haunt us some day.

* Okay, yes, and atmosphere and boats to Genoa and all the other things medieval Genoese couldn't survive without.

Cities are mortality sinks. The death rates, especially in the post-Black Death European cities for which we have the most evidence, are clearly outrunning the birth rates. But a quick shuffle of the economic order and they then turn that demographic frown right back upside down by attracting new residents, drawn by the prospect of participating in a waged labour economy. But as we've seen, that work can kill you, break down your bones and leave scars on your skeleton. What's more, it's done in a dirty, filthy world, where the sheer number of people leaves the streets paved in what I will euphemistically call horse-gold, and worse. Yet you cannot kill so many people and have it go unnoticed, and this may be the secret of the modern city's success. So many eyes on so many problems will, however grindingly slowly, force action. We saw it happen in an instant with Snow getting that pump handle removed,* but we saw it happening slowly, around the edges of the main thrust of our story, too. Leprosaria. Hospitals. Even the dreaded workhouses, which, remember, were not actually meant to kill people. The outrage at the working conditions, the living conditions and the *filth* of the city is funnelled into a force for change.

This, I think, is the heart of the matter. To strip away any nuance and come back to a simple idea: societies adapt – and physically concentrated urban societies perhaps more rapidly than most. It's easy to see the scope of the challenge we have set ourselves to adapt to. In the Marxist sense, cities remove people from the means of production. They accumulate people who must specialise in some form of exchange in order to survive. The power any individual has over this set-up is not some God-given equal share of an esoteric money pie, but instead reflects hoarded wealth (whether investment in physical survival, material goods or otherwise) and social status, both of which can be either built up or lost on the throw of dice.

* Turns out he did know something after all.

Cities allow the accumulation of surplus, both of people and things, and that can be exploited. That surplus becomes a more fluid object than the harvest it derives from; it can be invested, capitalised and used to start sending out little urban tentacles of power to feed back into the gaping urban acquisitive maw. The absolute necessity of having lapis lazuli beads brought a couple thousand miles from Afghanistan to a small river valley in southeastern Turkey, as happened at Başur Höyük in the third millennium, is not because some vain chief decided he needed a bit of blue to his bling (although that may have been part of the process). The lapis in those graves is the eventual outcome of a larger metals trade, a peoples trade and those Uruk Starbucks-style colonies. The push factor comes from the need to be seen with the power to acquire such an exotic thing in an increasingly settled, urban and urbane world that is filling up with such goods.

The urban tentacles, eventually, get everywhere. They encircle agricultural systems, locking people onto land and creating pools of labour (voluntary or not) that can be exploited to feed more profitable industries – whether it's Mesopotamian weaving factories or armies to besiege Naples. The global connections we have talked about in terms of the transmission of disease could just as easily be discussed as the transmission of a creeping network of trade, of exploitation, of shipping thing A to place B, and unintentionally creating a world of cities that are linked through roads and sea lanes, air travel, migration and, above all, the buying, selling and trading of things. As a species, we seem to have accidentally created a global market that we don't quite know what to do with. In its very bones and with all of its health consequences, the urban European world of the last few chapters is one only possible with lopsided trade, crass exploitation of labour and gross inequality.

And yet. I will rather controversially say that cities are not inherently a bad thing to adapt to. Inequality is a bad thing. Inequality, as we have seen through the course of this book, is the driving force in killing humans since we came up with this whole urbanism lark. And the driving force behind that inequality is two-fold. On the one hand, we have a world where the financial engine of all of our great cities is built on rewarding acquisition and hoarding capital; this is why the insanely wealthy own everything and you do not. Think what it takes for that statement to be true. The wealthy are not *better* than you, and many of them merely have better inherited investments. Those investments are crystallised in a monetary economy geared towards rewarding capital, and expressly for establishing fluid and expansive modes of financing things – things like expeditions to new continents where more money and more things can be acquired. New continents full of money are required because it's very expensive to be a nation-state – to field a standing army, to defend territory against other nation-states and, critically, to show off your extraordinary wealth to the other five people at the top of the urban food chain. None of this capitalisation would have been necessary if a little bit of lapis lazuli had been enough to wow the guys back home at Başur Höyük, but of course, once the transport is good enough, who *doesn't* have some foreign goods? No one in their right mind would go to the time and effort of acquiring endless amounts of bronze, trading it and developing complicated burial practices just for a bit of bling if there was an easier way to achieve power, the guys sitting at the edge of urban life just said: 'Hey. Great job on the Creation Eagle story. We'll bury you with a lot of tortoises.'

Probably not. Which is what takes us all the way around the houses to the second reason that it's *inequality* we should be blaming for urban deaths. To go right back to the example of Barbara and that Creation Eagle story in Chapter 5, collectivism is based on a way of life that depends on biological determinism including factors such as overall strength and the ability to give birth; and it sucks for a large number of people. What you call meritocratic your wives, children and pets might call tyranny. The past is not, as the yoghurt-crocheters would have it, an egalitarian wonderland. It may have been less brutal than the economic inequalities we can see written in our paycheques, in world poverty statistics, in the endless stream of sexist, racist, homophobic invective that is the comment section of any major news site and, most starkly, in the global health outcomes for the haves and have-nots. We cannot, however, assume that a world where the demands of biology were unsurmountable was uniformly kind.

Cities remove people from the means of production, and that cuts both ways. It allows them to support specialists, it allows for the creation of new kinds of specialisation and it keeps churning over the social structures that encourage innovation and adaptation. For a very large part of our urban history, the price cities have paid for that adaptability has been high. Specialists and non-specialists alike get wiped out in plagues that ferment in cities and run riot along the roads between them. Cities with walls get those walls smashed and their inhabitants perish alongside. That separation from the agricultural base for urban life means that, in a bad harvest year, neither love nor money will save the urban citizen. It has been a tortuous road, and I don't think that even the most optimistic of souls can say our urban future is one of immediate escape from the inequality that has plagued us for thousands and thousands of years. But those specialisations, as limited as they are to a privileged urban class, are freeing in a way that nothing else in our evolutionary history has ever been.

It's a very unique and recent specialisation, coming on the back of a long line of increasing specialisations in medicine, in health and in governance, that leads us to a reason for optimism about our urban future. These are the august international bodies, consolidated out of very specialist knowledge indeed, designed to compile and collate health policy and research from all corners of the globe. Their very existence is something that not even John Snow could have predicted. From our little isolated villages we have built ourselves into cities, and then linked those cities so thoroughly together that we come at last to a point where we can directly discuss health in a global context. Indeed, it is critical that we do so, because what kills people in China today can kill people in Brazil tomorrow; our world is so relentlessly connected that individual outbreaks of famine, war and disease are no longer the concerns of one great city or one nation-state, but the problem of all. At the time of writing, this is most evident in the collapse of Syria, which has sent refugees scrambling for security in an increasingly xenophobic, protectionist world. Two hours from the Syrian–Iraqi border, I have won (and lost) rather unequal footraces with children whose parents escaped from Syria into the uncertain limbo of Turkish asylum in the last few years and now live in a converted school behind the dig house. Five minutes from my home in London, I have watched the tents of refugees spring up among the cowslip and bramble that line the edges of the local park. These are global problems, which, while heartbreaking and seemingly intractable, offer at least some potential rays of hope.

The World Health Organization is one of those rays of hope, as are the many international charitable and government organisations that take world health as their remit. Our global world has its global challenges, yes, but it also has the seeds of the structures H. G. Wells and all those wide-eyed internationalists dreamed of more than a century ago – structures like the United Nations, the International Criminal Court and the World Health Organization. And it's through these international structures that we start to see the edges of a picture of human evolution that has yet to be fully painted in, the edges of new social and cultural adaptations that will change us yet again. We can see in the archaeology of our ancestors the signs of violent conflict, the burden of disease and the risk of being (or becoming) human. We can see in the archaeology of our early experiments with settling down the new threats we unleashed. We dig up our earliest cities and read the history of our urban past, with its wars and its diseases of density and inequality. We can look to the beginning of our own global era and see our cities as engines driving disease and conflict across continents and oceans, killing more people than even existed when we started this game 15,000 or so years ago. And finally, finally, we can look at the present, with these global risks, and we can see what urbanism has done to us.

The World Health Organization has drawn in the borders of our urban world with stark warnings. Urban living can be deadly, and this is as true now as it was 3,000 years ago. The urban poor living in 2015 are not only worse off than their richer neighbours, they are worse off than their rural counterparts. But by 2030 it is estimated that 60 per cent of the world's populations will be living in cities; we apparently are not giving up on this urban experiment any time soon. Is this nothing more than an exercise in masochism? Walking between the rough sleepers on a street in Cairo, Tokyo or San Francisco, you would be forgiven for believing exactly that. But there are moves afoot to break the crippling hold that inequality has on our species' health and survival. And, like modern urban cities, they are global in nature. Campaigns to eradicate disease such as polio and malaria cross national and social borders. Across the planet, individual cities subscribe to a global ideal of reducing health inequalities. There are targets identified: universal health coverage; using population density to combat disease transmission and establish immunity (instead of the other way around); treating those conditions that accompany urban life and inequality, like malnutrition; safe water, safe travel, safe homes; sustainability; and mobility. These are actually the headings given in the World Health Organization and the UN-Habitat's joint 2015 Global Report on Urban Health. And slowly, grindingly slowly, we are making progress. It can seem, in late 2016 as I write this, that the mechanisms of democracy and progress have broken down. The gears of our world have slipped. But it is still possible that among the gnashing metal teeth of our social structures, a new, better fit will be found. More and more, cities are leading the way in this. We have not reached the point where we are any *good* at living in cities, but give us time. It took hundreds of thousands of years to work ourselves up to settling down; but in 15,000 years we have transformed into a species where no aspect of life can be assumed untouched by urbanism. If we are going to be 60 per cent urban by 2030, then we are going to have to get better at it. Happily, adapting is what humans do.

Like nearly two centuries of wide-eyed speculators on the human future before me, I still see the city as a promise. It's currently an unfulfilled promise until we can deal with the inequality that has picked us off huddled mass by huddled mass, but our densely packed urban lives offer the opportunity to adapt, to invent and to collectivise in ways that don't rely on base biology. You would think it would be very difficult to write an entire book on such a long stretch of horrific human experiences and come out of it an optimist, but by and large I am. Inequality is the killer. Urbanism offers another way out. I can only really speak to my own experience, but of course that is one of the secrets of archaeology – interpretation is never really free of the personal. The city, you see, has jobs. The city thinks it's a good idea to provide specialist employment for female people who are adamant that thin-sectioning 10,000-year-old teeth is critical to understanding the human condition. So it seems that, however much it has warped our bones, rotted our teeth and inflamed our sinuses – the city is right.*

---

* A list of references is available at https://figshare.com/articles/BOB_full_bibliograpphy_pdf/4106247. DOI: 10.6084/m9.figshare.4106247.v1

ACKNOWLEDGEMENTS

Some of My Friends

I owe an immense debt of gratitude to far too many people to ever successfully list here, but I would very much like to have a go. My eternal gratitude to Jim and Anna, editors extraordinaire, without whose tequila shots (Jim) and bravery in the face of my unreasonable demands (Anna) this book would never have been written; and to Myriam for ensuring this book has all of the right words, all in the right order. It's a backhanded sort of thank-you given how I turned out, but I'd like to particularly thank all of the supervisors and advisers that have made me the archaeologist I am – particularly Simon Hillson, Dan Antoine, Louise Humphrey, Chris Stringer and Chris Dean, all of whom I very much hope will still speak to me even if I made them spit their coffee in various sections of this book. I am also thankful to my amazing colleagues, who have put up with my ranting about 'the book' in every possible circumstance – from Ros Wallduck and Ali Freyne in the Natural History Museum, to Laura Buck in just about every research institute in Japan. Enormous thanks are due to all of my amazing Team Anthro – Sarah, Nil, Izzettin, Oznur, Didem ('Didoo'), Martina, Fatma, Habip, Meltem, Metin (honorary) and Ihsan (honorary) – thank you for being so good at digging that I actually managed to get some of this book written on fieldwork! Thank you to Veysel for finding me the best place to work outside the British Library, and to Haluk Sağlamtimur for being one very patient excavation director.

This book would never have been possible without the help of friends and colleagues to answer even the most inane questions, and special mention goes to Güneş Duru for the most time spent explaining the Anatolian Neolithic. Thank you so much to Mihriban Özbaşaran and the rest of the team at Aşıklı Höyük for giving me so much to write about. An absolutely heartfelt thank-you to my early readers and commentators, who are largely responsible for this book (if it's good) and in no way at fault (if it's not). Thank you to Martha, to Hana, to Carolyn, to David and of course, to my mother. Thank you to my special academic advisers: Chris S. for trying his best to get me to understand human evolution, Rachel I. for sorting out the vitamin deficiency alphabet soup, Steve and Jess for explaining London archaeology even though I'm not really a 'real' archaeologist, Tim for telling me the difference between policy and politics in German, to Rachel F. for the Jon Snow jokes, and to Patrick for making sure I got Raquel Welch's bikini right. Thank you to the TrowelBlazers women who are my personal heroes (Suzie! Becky! Tori!), the Cavendish Road Yakut Appreciation Society, the SLC, my personal Reggie, and everyone who bought me a beer while I wrote this. Thank you to my family – Jenny, Kevin, Kaitlin, Allen, Lisa, Lynne, Doris and Connie – for understanding why I never come home. I told you I was busy.

There is one person whom I could never thank enough, who has read every word in this book twice over (and added a 'u'in like half the words). This book would never have been possible without the tireless support of Andy Bevan, to whom I owe a huge debt of gratitude and probably a drink or two.

Index

Aboriginals 104, 130
Abu Hureyra, Syria 46–47, 48, 253, 257
accidents 266–68
Ache people, Paraguay 27
Africa 74, 82, 93, 117, 139, 192, 203, 250
Agricultural Revolution 46
agriculture 41–47, 52–53, 93
  sedentism 78–80, 85–86
air pollution 290–94
alastrim 212, 216
Alexandria, Egypt 289–90
Americas 28, 82, 85–86, 93, 139, 203
  Mesoamerica 160–62
  smallpox 215–16, 223–24
anaemias 108–109, 112, 213
Anatolia 44–45, 53–54, 78, 81–83, 87–88, 100, 117, 162–63, 215, 221, 254–55, 301
animal pollution 287–88, 290
Antikythera, Greece 57–59, 75–76, 151, 282
archaeobotany 39–40, 42, 45, 255
armies 188
arthritis 47
  osteoarthritis 261–65, 269, 273
  syphilitic 234
Asia 82, 117, 139, 201, 203
Aşiklı Höyük, Turkey 17, 53–55, 82, 87, 90, 100, 254–55, 292, 301
  domesticated animals 66–67, 71
Atapuerca, Spain 112, 119, 174
Atlit-Yam, Israel 202
Australia 78, 104, 130, 137, 215
Australopithecus africanus 68–70
Aztecs 160–62, 171

babies 26–27
Barker, David 102–103
Başur Höyük, Mesopotamia 163, 165–66, 260, 307
Bedouin 15–19, 32, 253
bejel 231–34
Belyaev, Dmitry 60–61, 65
Bills of Mortality, London 220–21, 283–86
Binford, Lewis 27–28, 98
bioarchaeology 11–14, 29–30, 52–56, 81, 85, 104–109
biological warfare 221–23
Black Death 209–10, 217–23, 224, 226–27, 250–51, 281–82, 295–96
Boccaccio, Giovanni 218–19
Bocquet-Appel, Jean-Pierre 25–27, 41, 55, 86
bones 11–12, 33–34, 299
  Abu Hureyra 46–47, 48
  ageing 260–61
  bone scans 258–59
  dogs 62–65
  domesticated animals 61–62
  malnutrition 105
  metabolic disease 87, 105, 113–15, 200–201, 213, 261
  osteoarthritis 261–64
  syphilitic lesions 233–35
  tubercular lesions 200–201
British Institute, Ankara 54
Bronze Age 57, 70, 138–40, 142, 171, 178, 181, 201, 260
brucellosis 68–70, 75–76
bubonic plague 191, 209–10, 219–23
Bunon, Robert 2401–41, 244
burials 299–300
  deviant burials 150–54

Caffa, Silk Road (Fedosia, Ukraine) 221–22, 231, 304–305
Cahokia Mound Builders, Mississippi 18

cannibalism 120, 174–75
Çatalhöyük 53–54, 82, 87–91, 100, 292, 301
cats 65–66
cattle 53, 68, 74, 181
cereal crops 42–45, 106
Chalcolithic 112, 260
Chasko, W. J. 27–28
child abuse 144–47
Childe, V. Gordon 40–45, 78, 84, 94–95, 164, 269, 300
childhood illness 52–53
chimpanzees 10, 26–27, 173–74, 192
China 28, 138–39, 148, 150, 201
cholera 29, 211, 294–7
Chumash culture 85, 98, 120–25, 137, 167, 190
cities 9–10, 14, 94–95, 115–16, 304–305, 311–12
  disease 189–90, 305–306, 309–11
  livestock 289–90
  plagues 226–27
  syphilis 230–31, 150–51
  violence 117–19, 160
class structures 95–98
Columbus, Christopher 236, 238, 244, 246–47, 250
communal settlements 100–101
Cooper, Jago 85–86, 98
coronavirus 206
Council for British Research 15
cradleboards 277–78
Crete, Greece 57–59
Cucuteni-Trypillian culture, Ukraine 95
Cultural Revolution, China 271–72

Darwin, Charles 60
demography 41
Denisovans 82, 84, 191
Despenser, Hugh 157–59
diabetes 11, 36, 37
Dickson Mounds, Illinois 46, 52, 87
diet 35–36
  palaeo-diet 10, 37–39, 106
disease 9, 29, 188
  animal crossover infections 70–71, 190–91, 192–93
  domesticated animals 66
  sedentism 29, 191
  see cities
DNA, ancient (aDNA) 31, 55–56, 80–83, 191–92
dogs 62–65, 66
domesticated animals 33, 55, 57
  Aşıklı Höyük 66–67, 71, 288, 290
  brucellosis 68–70, 75–76
  cats 65–66
  crossover infections 70–71
  dogs 62, 64–65
  goats 58–60
  livestock 67–68, 288, 289
  milk 74–75
  pet-keeping 62–64
  Secondary Products Revolution 71–73
  tameness 60–61, 67
Dunbar, Robin 99–100
Duru, Güneş 17, 54

Ebola virus 191, 206
elder mortality 23, 24, 25
epidemics 71, 192, 197, 203–207, 209–11, 221–23
  Black Death 209–10, 217–23, 224, 226–27, 250–51, 281–82, 295–96
  cholera 294–97
Erdal, Yılmaz and Dilek 54
Esin, Ufuk 53
Europe 40, 74, 76, 81–82, 93, 117, 201
  Black Death 219–21, 226–27, 250–51
  syphilis 236–40, 243–47, 250
executions 155–59

faces 49
farming 37, 40, 55–56, 254
  accidents 267–68
  Urban Revolution 98–101

female fertility 25–26, 86–87
Fertile Crescent 44–45, 78, 83
Finlayson, Bill 15–18
fortifications 173, 180–82
Foucault, Michel 172–73
foxes 60–61
Francisella tularensis 215, 221
Fuller, Dorian 40, 43

Galtung, Johan 143
Glasgow 112–13
goats 58–60, 66–67
Göbekli Tepe 90, 94, 271
greenstick fractures 131–32
grinding grain 20, 43, 47–48, 257
growth disruptions 31, 54–55, 80, 87, 104–105, 244, 272
Guevara, Che 193

Hadza people 38–39, 254
hantavirus 191
Happisburgh, Norfolk 280–81
Harappa, Punjab 141, 201
head injuries 124–29
height 12, 30–32, 34, 102–104, 300
  astronauts 261
Helicobacter pylori 192
Herculaneum 70
Hilazon Tachtit, Israel 95–96
Hillson, Simon 49, 244
hippies 170–71
Hittites 215, 221–22
HIV 230, 239, 247
Hodder, Ian 88
hominins 10, 25, 97, 300
  Australopithecus africanus 68–70
  cannibalism 174–75
  Happisburgh, Norfolk 280–81
  Helicobacter pylori 192
  Homo naledi 264–65
  Sima de los Huesos 119–20, 174
Homo antecessor 174
  erectus 80, 107–108, 201
  naledi 264–65
Hudson, E. H. 247–49
human sacrifice 160–62, 165–68
hunter-gatherers 16, 20, 21
  Kalahari bushmen 38–39
  osteoarthritis 262–63
  skeletal measurements 29–32
  violence 139
  warring groups 176–77
Hunter, John 242–43
hypoplasia 52, 87

Iberomaurusian culture, Morocco 50
Incas 134–35, 160–61, 166
India 82, 195, 201, 203, 213–14, 215, 219–20, 293, 294–95
Industrial Revolution 14, 211, 269
inequality 101–103, 302, 307–309
infant mortality 23–24
influenza 29, 192
Institute of Archaeology, University College London 40
Institute of Cytology and Genetics, Siberia 60
Inuit 27–28, 105, 138, 262, 265
Irish Potato Famine 274–75

jaws 49
Jenner, Edward 214
Jericho, Israel 70, 159, 164, 180–81
Johnson, Samuel 198–99
Jomon culture, Japan 55
Jordan 15–19

Kalahari bushmen 38–39
Keeley, Lawrence 137–38, 173, 180–81, 184, 187–88, 303
Kellis 2, Dakhleh Oasis, Egypt 145
Kenyon, Kathleen 180–81
Koobi Fora 107

lactase 73–74, 76
Larsen, Clark Spencer 49, 55, 130
Late Woodland 48, 149
leprosy 191, 193–99, 204–206, 211
Levant 15, 20, 28–29, 45, 78, 83, 256–59
Lewis, Mary 145–46, 157–58, 293
life expectancy 22–3
  Palaeolithic 12

Linearbandkeramik (LBK) cultures 181–83
London 279–80
air pollution 290–91
Bills of Mortality 220–21, 283–86
Black Death 281–82, 295–96
hygiene 286–87
Roman occupation 281, 294
water pollution 294–97
work 282

maize 43–44, 85–86, 114–15, 149, 259, 275
malnutrition 9, 31, 52, 87, 103–106, 203–204, 213, 275, 301, 302, 311
Malta 75
Malthus, Thomas 188
Marxism 41
maternal energetics 25–28
Maury, Pierre 288–89
Maya 140, 160–61, 248–49, 260, 273
measles 29, 192
Mediterranean 58, 75–76
Eastern Mediterranean 20, 225, 232, 245
mercury 241–42, 269–70
Mesolithic 21, 31, 74, 134, 176–77
Mesopotamia 41, 73, 111–12, 117, 140, 162–64, 185, 273–74, 280, 303, 307
metabolic disease 87, 105, 113–15, 200–201, 213, 261
metallurgy 117, 139–40
miasma 287, 291
mice 20, 65, 191
milk 72–76
Mississippian cultures 18, 259
Moche 160–61, 187
Molleson, Theya 47, 257, 267
Mongolia 95, 138–39, 150
mortality 23–25, 29, 41
publication of records 220–21, 283–86
Mother Theresa 193
Mound Builders 18, 46
mummies 135–36, 161, 166, 201–202, 214–15
muscle activity markers 256–60
Musular 90
Mycobacterium 190, 193, 198, 204
africanum 203
bovis 200
leprae 194, 198, 205, 206
tuberculosis 199, 200, 201, 202–203, 205

Naqada culture, Mesopotamia 111–12
Nataruk, Kenya 179, 182
Natufian culture 20–21, 28, 31–32, 45–48, 67–68, 83, 85, 95, 259
Tortoise Queen 95–96, 100
Neanderthals 22, 39, 42, 82, 134, 191–92, 258–59, 265–66
Near East 28, 40, 41, 42, 52, 82, 93, 140, 248
Neolithic 25, 55–56, 171, 201
Neolithic Demographic Transition 27–32, 41, 48, 55, 202, 266, 305
Neolithic Revolution 40–46, 51, 54, 78–80, 93–94, 300–301
ancient DNA 80–83
slow pace of development 83–87

obesity 35–36
Ofnet Cave, Bavaria 177, 179
Olmecs 160
osteoarthritis 261–65, 269, 273
osteomyelitis 212–13, 232
osteoporosis 132
Ötzi the Iceman 135–36, 139, 141, 176, 184
Özbaşaran, Mihriban 17, 53–54

palaeo-diet 10, 37–39, 106
Paleo Diet™ 36, 106
palaeodemography 22–23, 25–25, 272
Palaeolithic lives 11–14, 22
life expectancy 12

pandemics 29, 206, 210, 221, 223–26
Papua New Guinea 93, 192
parasites 57, 70–71, 108, 115, 145, 193
parry fractures 129–31
pellagra 114–15
Pepys, Samuel 220, 284
phosphorus 270–71
pigs 68, 69,74
pinta 231–32, 235, 237, 246
Pinzón, Martín Alonso 237
plagues 29, 191, 207, 210–11, 225–27
  Black Death 209–10, 217–23, 224, 226–27, 250–51, 281–82, 295–96
  smallpox 211–17
plant-based foods 37, 42–45, 93
pollution 287–88, 290
  air pollution 290–94
  water pollution 294–97
Pontecagno, Italy 139
Pontiac people 223–24
population 28–29, 41, 46, 188
potatoes 43–44
Pre-Pottery Neolithic 21, 100
pregnancy 26
Pueblo cultures, ancient 149–50, 178–79, 266–67, 272–73

rats 66, 219, 220
Rawcliffe, Carole 197, 205–206
rice 37, 42–44, 55, 85, 93
rickets 110–13, 234, 284
Robb, John 138–39, 142, 175, 178, 183, 187
Rome, ancient 207, 215, 226
root crops 43–44
Rwanda 183

Salğlamtimur, Haluk 163–64
San people 38–39, 138, 148–49
Scott, James 99
scrofula 198–200, 284
scurvy 109–12, 275, 284
Secondary Products Revolution 71–73
sedentism 15–16, 18, 21, 25–34, 40–41, 68, 74, 176, 266
  agriculture 78–80, 85–86
  development 93–95, 149
  disease 29, 191
Shangshan, Yangtze 18
Shanidar Caves, Iraq 265, 266
sheep 66–67, 289–90
Shennan, Steven 56, 97–98
Sherratt, Andrew 72
Silk Road 220, 231, 304–305
Sima de los Huesos, Spain 119–20, 174
sinusitis 292–94
skeletal measurements 29–32
slavery 272–73
  North America 115, 272
smallpox 202, 211–17, 223–24, 239–40
Snow, John 296–97, 306
Spitalfields Project, London 263–64, 268
Staphylococcus aureus 213
starvation 102–103
status 95–98
Stele of Vultures 185
Sumeria 273–74
syphilis 211, 225, 229–33
syphilis 211, 225, 229–34
  arrival in Europe 236–40, 243–47, 250
  congenital syphilis 234–35
  Hunter, John 242–43
  mercury treatment 241–42
  origins 247–51
  Salpêtrière, Paris 240–41
  types 235–36
Syria 16, 33, 46–47, 215, 260, 303, 310

teeth 12, 33–34, 47–48, 68
  congenital syphilis 235, 240–41, 244
  domesticated animals 61–62
  malnutrition 104
  tooth decay 49–51
  tooth formation 51–52

teeth lines 52–53
Aşiklı Höyük 54–55, 87, 301
Tell Brak, Syria 260
Temple, Dan 55
Tollensee, Germany 178, 179, 183
Toxoplasma gondii 66
Treponema pallidum 231–33, 235, 243–50
Trinkaus, Erik 134, 265
tuberculosis 199–207, 211
Turkana, East Africa 138, 176–77
typhoid 29, 215, 295

Überlingen, Germany 224–25
Ungar, Peter 48
urban life 9–10, 14
Urban Revolution 40, 94–95, 300
class structures 95–98
farming 98–99
inequality 101–103
social structures 99–101
Uruk, Mesopotamia 162–65, 184, 307
US Centers for Disease Control and Prevention (CDC) 69, 199, 230

Variola major 212, 215–16
minor 212, 216
Vibrio cholera 294, 297
violence 117–19, 302–303
Chumash 120–23, 125–26, 137
deviant burials 150–54
executions 155–59
head injuries 124–29
human sacrifice 160–62, 165–68
interpersonal violence in the ancient world 137–42
Ötzi the Iceman 135–36
parry fractures 129–31
sharp force trauma 132–35
Sima de los Huesos 119–20, 174
violence against children 144–47
violence against groups 150
violence against women 147–50, 303–304
witches 154–55
vitamins 104–12, 264–65
Von Clausewitz, Carl Philipp Gottfried 172–73, 179
Von Cramon-Taubadel, Noreen 29–30

wadis 18–22
Wadi Arabah 32
Wadi Faynan 19–20, 21, 32
Waldron, Tony 23, 262, 268
Walker, Phil 108–109, 121–22, 124–25, 134, 137, 145–46, 148–49
war 142, 169–74
archaeological evidence 175–84
early war memorials 184–85
modern warfare 186–88
participants 185–86
water 18
water pollution 294–97
weapons 128–30, 132–34, 139, 169, 175–76, 178, 183–84
witches 154–55
women, violence against 147–50, 303–304
work 253–54, 306
Industrial Revolution 269
modern Egypt 275–77
muscle activity markers 256–60, 272, 273
osteoarthritis 261–64
Sumeria 273–74
workhouses 274, 306
World Archaeological Congress 15
World Health Organization 24, 103, 148, 199, 310–11

Yangtze cultures 18, 43, 203
yaws 231–35, 245–48
Yayoi culture, Japan 55
Yersinia pestis 210, 220–21
Younger Dryas 21, 31

Zagros Mountains culture 82–83
zoonoses 68, 71
Zuk, Marlene 13, 36